THE MUCKELMAN PARADOX

SERIES TWO–TRUTH AND UNEXPECTED CONSEQUENCES

H. Gary Apoian

The Muckelman Paradox is a novel. Other than recorded historical facts, the characters and incidents are fictional.

Table of Contents

Acknowledgment

I would like to thank Amazon Publishing Company, particularly Mathew Johnson and his team, who always made themselves available to the writer with wonderful advice and patient assistance.

About the Author

The author has dreamed of becoming published since his undergraduate and law school studies at Northwestern University. After retiring from his law practice, that dream became a reality.

Never a quitter, he persevered through what seemed to be the insurmountable challenges of trial practice. The author has been gratefully acknowledged by Martindale-Hubbell as The Client Champion in 2019 and 2020 and has been honored to receive The Lawyers of Distinction designation for excellence in civil litigation.

INTRODUCTION

The Muckelman Paradox Series is a comical journey where laughter and deep thought collide. The characters take the reader for a wild ride through the universe of humor and existential thought.

The Series' philosophical undertone turns parallel paths of reality and delusion into the realm of contemplation. Readers will ponder the more profound meaning beneath the humor.

Various plot twists transcend expected boundaries, both literally and metaphorically.

The Muckelman Paradox Series is a wild, whimsical, and wonderfully bizarre experience that will leave readers laughing, thinking, and eagerly anticipating the next twist in the cosmic comedy.

Chapter 1:

New Friends

Todd Montgomery and Elise Manchester met while they were students at Brigham Young University. They both majored in political science. After graduation in 2008, the twenty-three-year-old couple married and moved from Provo, Utah, to their hometown of Salt Lake City. The return home provided the comfort of family and close friends active in the Mormon community.

After graduation, Todd changed his plans and decided not to pursue law school. He liked the law, but being a staid lawyer did not appeal to him. Making more money practicing law was never an issue after his sizable inheritance from his uncle, William Montgomery. Uncle Bill never married and earned his sizable wealth as a franchisee of several chain restaurants.

Instead of going to law school, Todd applied to the FBI. He entered their twenty-week training academy in Quantico, Virginia. He graduated at the top of his class and received his first choice for an open position at the Salt Lake City FBI office. In July 2009, the Bureau promoted him to the new Director of the Manhattan District Office.

Todd drove a new Silver Rolls Royce Phantom. Elise felt uncomfortable riding in such a flashy car. She preferred her red 2008 Mustang G.T. Because of the FBI's financial disclosure rules, everyone at the Agency knew the source of Todd's wealth, but that did not prevent fellow agents from teasing that he was on the take when they saw him in the Rolls. Bystanders gawked at the luxury car whenever he appeared with its oscillating red light and siren. The young women in the crowd gasped in awe as the gorgeous hunk exited the exotic car.

Elise and Todd were a handsome couple. They fit the perfect Hollywood template of a fit, young, and loving couple—they resembled a traditional stereotype of the working husband and homemaker wife. When Todd arrived home in the evening, Elise always greeted him at the door with a smile and a kiss, wearing one of her frilly dresses with a single strand of pearls. They sat in the study and discussed their day over cocktails before dinner.

***

Although the couple missed Salt Lake City and experienced a brief culture shock of New York City's constant noise, fast pace, and cost of living, they adapted. The couple fell in love with the newly gentrified Westside of New York. The Montgomerys did not have children or plan to have any soon. However, they bought a four-bedroom Cooperative at 550 West 45th Street, at the corner of 10th Avenue and 45th. Their luxury apartment on the building's top floor spared them from the constant racket of New York City. They loved Manhattan but found the City a difficult place to make the type of friends they sought.

***

During an early autumn stroll, their loneliness faded when they visited Mr. Biggs' Restaurant and Bar. An FBI colleague of Todd's recommended the bar and grill in the old Hell's Kitchen Neighborhood at 9th Avenue and 43rd Street. The bar and eatery earned the reputation for serving the best hot wings and the largest selection of imported draft beer in Midtown Manhattan. The crowded bar and grill confirmed its popularity. Rather than wait an hour for a table, the couple sat at the bar. Todd sat beside a beautifully dressed woman with a small dog on her lap.

Chapter 2:
The Small Dog on My Lap

The couple introduced themselves to Cherie Egegian. Her Pomeranian, Mucky, was hungry and in no mood for idle chitchat. "Gucci girl, just order. My stomach is growling." Nearly one month had passed since Cher adopted Mucky from an animal shelter in St. Louis, Missouri.

***

When she first heard the animal speak, Cher feared she had suffered a delusional event. The dog claimed to be the bizarre transformation of Irving L. Muckelman, a seventy-seven-year-old ornery man who worked in New York's Garment District. Only she and two close friends, George Markarian and his paramour Maria Allison, could hear Mucky's confused and illogical words of wisdom. Mucky's transformation, rather than death, followed a commercial airline accident with no survivors. The plane's manifest listed 189 crew and passengers, but 188 bodies were recovered. Presumably, the discrepancy accounted for Irving Muckelman.

Mr. Muckelman had traveled from Manhattan to Florida for his brother's funeral and stayed at the Longboat Key home of George Markarian and his live-in girlfriend, Maria Allison. He spent the last day in Longboat Key, enjoying the luxurious swimming pool with Maria's black cat, Onyx. Mr.

Muckelman and the cat shared a common trait. They did not mingle with people. But the cat broke with custom and shadowed Mr. Muckelman the entire day.

When Maria arrived home from her car dealership, Mucky was asleep on a chaise lounge with Onyx playing sentry while perched on his chest. The nine-year-old cat had previously repeated this contrary behavior to three people: his original owner and Maria's brother and father. Each unexpectedly died the following day. Maria worried Mr. Muckelman would become the fourth victim. When Maria called George at the office about the unsettling story, he discounted the cat's horrific psychic prediction.

That evening, the three left for Sarasota International Airport for Mr. Muckelman's 8:15 p.m. departure for New York. When the couple returned home, they watched television with Onyx nestled on the back of the couch. A–bulletin interrupted programming with breaking news. Mr. Muckelman's airplane had crashed after take-off. Although the government's recovery team never found Mucky's body, the Sarasota County Medical Examiner issued his Certificate of Accidental Death on October 29, 2009.

The NTSB could never recover or explain the missing body of Irving L. Muckelman because he had been mystically relocated to a St. Louis, Missouri, animal shelter and transformed into a caged Pomeranian. He could talk but had no audience because he could not be heard.

The college student from St. Louis, now sitting at the counter of Mr. Big's Restaurant and Bar, adopted the dog, intending to begin her college career with a canine roommate. Cherie

Egegian slowly acquired the ability to hear her new pet speak. She learned his name, Irving L. Muckelman, and his lifetime occupation as a lead floor salesman and tailor at a shop in the Manhattan Clothing District.

The seventy-seven-year-old man, now a Pomeranian, lived up to his predecessor's reputation with the behavior of an abrasive and annoying animal. He escaped death in the tragic airplane accident because his Maker commanded his atonement. The Boss saw goodness buried deep within Mr. Muckelman's soul and allowed him a final opportunity to redeem his earthly misbehavior instead of receiving a direct ticket to a place where stylistic pitchforks were fashion accessories. HE also chose Cherie Egegian as his conduit to facilitate Mucky's redemption.

After Cher overcame her feared psychosis, she reluctantly functioned as Mucky's medium for communicating with and convincing George Markarian and Maria that the Pomeranian was his longtime acquaintance and the brother of his deceased partner. Cher and Mucky accomplished their objective after a long train ride from St. Louis to Sarasota. Three people who believed in Mucky's celestial experience, Cher, George, and his true love, Maria, could hear and freely communicate with him.

***

Cher ignored Mucky's whining while Elise admired her dog; Mucky demanded that Cher explain their unusual situation. "Tell them who I am and what happened." Cher had no intention of informing her new acquaintances of the unbelievable details of her odyssey. Mucky objected to the snub. "Okay, disregard me at your own peril."

Todd complimented Cher's style of dress. She thanked him and said, "I'm delighted you like the look. I am studying fashion design at the New York Fashion Design Institute and created the dress. I'm in the last year of my foundation grant. My stipend and housing allowance will end when I graduate, and my life in the real world begins."

Elise explained why she and Todd left Utah. "My husband worked for the FBI in Salt Lake City. After being with the Agency for over a year, the Bureau promoted Todd to the Director of the New York City FBI Field Office."

Cher noticed the initials CTRW on the couples' simple wedding bands and asked, "What do the initials signify?" Todd explained they related to their Mormon Faith and meant: Choose the Right Way. Todd explained, "We're devout Mormons and miss the strong spiritual influence of Utah. There aren't many Mormons like us in Manhattan, particularly Fundamentalists. Please understand we don't wear our religion on our sleeves. We seek friendships with people who live by the sentiment expressed by the initials on our rings."

Cher asked Todd to explain the difference between a Mormon and a Fundamentalist. He glanced at his wife and evaded a direct answer. "We have to be careful how we explain that. Once we know you better, we'll be more specific. For now, I'll say it's more of a lifestyle difference in daily living." After finishing their pleasant lunch, they exchanged addresses and phone numbers while Mucky gobbled the last of his burger. They agreed to meet soon.

***

After returning home, Mucky and Cher watched a movie on television. Mucky continued distracting her with his disturbing habit of talking during the show. "I liked those folks we met today. They're down to earth, unlike this town's four-flushers. We should tell them about that foreign element who just moved across the hall. That guy doesn't seem right. I like having the head man at the BFI as a friend."

Cher said, "Please be quiet. He won't be your friend for long if you call him a garbage collector. Todd is with the FBI, not the BFI." Cher gave up following the movie and asked, "Besides, what's wrong with our new neighbor?" Mucky offered an extensive list of criticisms.

"He looks like a terrorist. I always see him in the mail room picking up boxes stamped: CHEMICALS ENCLOSED HANDLE WITH CARE. I hear his door open and close in the middle of the night, and whoever it is doesn't stay long. There are plenty of those visits. He's a loner who doesn't smile or greet anyone he passes in the building. He looks mean and angry." Mucky's description of the man across the hall caused Cher to say, "I'll think about it and decide if we should mention it to Todd or if I should call George and Maria and see what they think." Mucky disagreed about involving George, "He might be out of his league here. No question, he's a shrewd businessman, but the head guy at the

Manhattan FBI is what we need. Compared to Todd, George's arsenal is a spitball."

***

The mystery neighbor, Jon Malik, a Muslim with a work visa, flew on a one-way ticket paid with cash from underground Muslim extremists. He lived in the war-ravaged Capital City of Sarajevo, Yugoslavia, the 1980 site of the Winter Olympics. War radically changed the beauty of the snow-covered venue. Yugoslavia collapsed in 1992 from the attacking Serbian forces. The forty-two-year-old Malik, a mortuary assistant in Sarajevo, changed his outlook on life after his mother and father perished in the ethnic cleansing when Serbian forces attacked Muslims in Eastern Bosnia. His apolitical existence turned fundamentally political. His job spared him from the wrath of war with an automatic military exemption. Morticians were needed as much as doctors for the combatants who lost their battle with death.

The man of mystery was tall and lanky. At six foot eight, he looked like he could play basketball for the New York Knicks. Malik's hardened looks and persona scared the children in the building. He kept company with one woman, but the rumor mongers claimed she was a prostitute. When she visited, they never left the apartment.

Chapter 3:

Nice to Meet You

The following morning, Cher decided to welcome her new neighbor to the building. When Mr. Malik opened his door, Cher stood holding a tray of cookies with Mucky at her side. "We wanted to introduce ourselves. We live across the hall. I baked you some cookies."

After Cher handed Malik the cookies, he nodded. She assumed the tenant didn't speak because of his poor English. Mucky's opinion of their new neighbor was clear and straightforward, and he used his usual manner of humorous expression. "This guy gives me the creeps. Ask if he's related to Boris Karloff." Cher asked the inexpressive man, "Do you understand what I'm saying?" Mucky slipped inside the apartment, and Mr. Malik was displeased. "Remove your animal from my unit!" His fluent English and behavior convinced Cher that Mucky correctly pegged their new neighbor.

***

Cher telephoned Todd's office. Still in the field, she left a message for him to call. After telephoning, Cher took the elevator to the parcel room and rummaged through the packages while Mucky played watchdog. She grabbed one of the packages addressed to her neighbor with the same chemical warning Mucky described. Cher shook the box, heard nothing, took pictures, and returned to her apartment.

Her brazen behavior of invading someone's privacy surprised Mucky. "You sure have changed, Gucci girl. You're now stealing from the mailroom." She corrected her friend. "Thanks, Perry Mason, but taking a picture is not stealing."

***

When Todd arrived at the office, fellow agent James Ryan peeked into the room. Todd smiled at him and said, "It must be a slow day. You look like you're about to fall asleep." Todd's secretary, Mary, broke into the conversation. "Excuse me, Mr. Montgomery, you received a message from Cher Egegian. Here's her number. She sounded like it was urgent." Todd rose and said, "Well, maybe business just picked up."

Agent James Ryan, the Assistant Director at the Manhattan Bureau, won many awards during his tenure at the FBI. He and Todd became good friends and often chatted about work and sports. "Boss, it may be important. I'll see you later. Let me know if you need me. You know I'm a pretty good profiler." Todd disagreed, "You're too modest, James— you're the best!"

Before Todd returned Cher's call, he spoke to his wife. "Elise, did you talk to Cher Egegian today? She called the office while I was out; Mary brought me her message and said it sounded important. I wanted to see if you knew what she might want. Okay, thanks; I'll call you later." Cher had not spoken to Elise since they met at Mr. Biggs.

When Todd returned Cher's phone call, she related her encounter with Mr. Malik and described the boxes of chemicals. Todd's alarm bells sounded. He put her on hold and buzzed James to come into his office. "James, I have Cher Egegian on the phone. A foreign national moved across the hall from her apartment. He receives boxes marked with chemical warnings. He doesn't sound like a tourist. He's probably here on an H-2B work visa. I'll check with the State Department. See how he paid for his ticket and if he booked a return flight. Also, let's run a complete background check."

He returned to the telephone. "Cher, I have Assistant Director James Ryan in my office. I have you on the speakerphone." Todd said, "We need to talk tonight because tomorrow I have a D.C. conference with the Bureau's Field Office Directors. I don't want to be seen by Malik at your place. Are you able to come to our house this evening around eight?" Cher agreed, and Todd instructed her to write a factual account of everything she remembered. He said, "You have our address. Bring the pictures, your narrative, and don't engage the guy."

Todd hung up, looked at James, and asked, "What do you think?" James, always willing to act sooner rather than later, said, "Hell, yes! It looks like we've got a live one. Should I notify Homeland Security?" Todd told him to stand down until he met with Cher. "He's presently a person of interest. We don't have sufficient actionable facts. I'll send you a teletype after our meeting. His name is Jon Malik. After you run his name, obtain his picture, and everything available from the Kennedy TSA, have the State Department contact our Serbian Embassy about his visa.

***

Thrilled when he learned about Cher's meeting with Todd, Mucky couldn't wait to see him. "I'm glad you took my advice. What time are we expected?" Cher said, "I'm going alone." Mucky didn't understand why. "Gucci girl, calling Todd was my idea. I should be there." Cher explained many people don't like dogs to visit with their owners. Mucky still objected. "Just tell them who I am." The debate continued. "If you think I'm going through that again, you're nuts. You're the meddler who put this into play." As Cher walked out the door, she said, "You better behave. If you act up, we'll play a new game—Microwave Wednesdays." Mucky went to the bedroom to sulk. Cher arrived at Todd's at 8:45 that evening.

Chapter 4:

The Battle between Right and Wrong

Cher apologized to Elise for arriving late. "Oh, Todd doesn't care. I'll walk you to his office. I don't know any details, but I saw him doodling after speaking to you. When he's considering something important, that's his way of thinking." The girls walked into the office, and Todd greeted Cher. Elise told them to get the bad guys and closed the six-inch door.

The office impressed Cher. "Your door blends into the wall." While she spoke, Todd grabbed a legal pad and said, "Since nine-eleven and the creation of the FBI Joint Terrorism Strike Force, the Office of Management and Budget requires homes of all FBI Directors to have the equivalent of a safe room; that explains the door. The OMB uses the same government contractor that fortified the White House Oval Office. All video conferencing and phone lines are secure, with direct hookups to the White House, the Pentagon, and the Defense and State Departments. You are sitting in the safest room in New York."

Todd asked, "Do you mind if I tape our conversation?" Cher agreed, and Todd continued, "Before we begin, may I see the photographs you mentioned on the phone?" While he examined the pictures, Cher provided additional background. "Mucky and I went across the hall to his apartment with cookies to welcome him to the building. He wasn't friendly; in fact, he looked angry. He nodded his head rather than speak. I assumed his English was poor, but he became upset when Mucky entered his apartment. Malik used perfect English with no accent when he ordered Mucky out."

Todd asked if Cher saw a computer or electronic equipment in the apartment. She said, "All Mucky saw was a laptop." Todd laughed and asked, "Oh really. How did he pass this information to you?" Cher realized what she said and explained her gaff well. "Todd, I'm a nervous wreck." He laughed and said, "I'd sure hate to tell DC I'm setting up an interview with your dog. We can find a treasure trove of information with his I.P. address and the seizure of his hard drive. Still, before we begin, we must have sufficient facts to satisfy probable cause of a crime before the Foreign Intelligence Court issues a search warrant.

"Our Assistant Director, James Ryan, is running a complete database on Malik. Homeland Security's TSA Division at Kennedy and our Bosnian Embassy are working to obtain copies of Malik's passport, visa history, and the purpose of his foreign travels."

Todd asked, "How did you enter the parcel room to take the pictures? Cher said, "All residents can access the parcel room by entering their apartment number and password on a keypad. The room has 24/7 surveillance." Todd asked, "Did you try to open or tamper with any boxes?" The question perplexed Cher. "No, but he continually receives these packages. I can open one if you want." Todd emphatically said, "Oh, God no! Then we'd have a serious problem getting a warrant.

"Let me explain. The U.S. Constitution protects U.S. citizens and foreign nationals in this country with the same rights to a reasonable or legitimate expectation of privacy. A law enforcement agency must have probable cause of a crime before invading that right. However, this requirement excludes private individuals acting on their own. But if I asked or suggested that you open a box, the Supreme Court has ruled you would lose the private search exception, and everything we learned afterward could not be used in Court. The U.S. Supreme Court has reasoned that the private party and the government have transformed the prior private status into an agency relationship between the private party and the government." Todd saw Cher's hand tremble and said, "We've been at this for an hour. Why don't we take a break."

***

Todd and Cher joined Elise in the living room. She remained a bit flushed and said, "I couldn't do this job without suffering a nervous breakdown." Elise wanted to change the discussion. "Enough of law and justice. Tell us about yourself." Cher omitted any mention of Mucky, the most intriguing part of her life. "I'm a St. Louis native attending a fashion school in New York. I have a few close friends. I like sports, watching a good movie, dining out, and reading—pretty boring stuff."

Cher hesitated but asked about their Mormon religion. "At Mr. Biggs, I found you guys very dedicated to your Church.

I wish I were as dedicated. I'm a member of the Armenian Orthodox Church but a regular absentee on Sundays.

Todd said, "Cher, there has been constant controversy about our faith. Mormons believe in the crucifixion, resurrection, and divinity of Jesus Christ. The difference is that Mormons claim that God sent more prophets after Jesus's death. Our founder, Joseph Smith, and his followers accepted the Bible and the sacred scriptures of the Mormons. Still, they diverged from orthodox Christianity, asserting that God did not exist in three distinct entities: The Father, Son, and Holy Spirit. Mormons also believe that faithful church members will inherit eternal life as gods. Other unique doctrines include the belief in pre-existing souls waiting to be born and in the salvation of the dead through retroactive baptism. The Church came under attack for its practice of polygamy between 1852 and 1890.

"Do you remember when we met, you asked about Fundamentalist Mormons?" Cher nodded in the affirmative. "Well, we must be careful who we tell because the Church of the Latter-day Saints does not recognize us. We could be excommunicated if our beliefs became known. Even worse, my career with the FBI would end because plural marriage, a form of polygamy, has been criminalized by the U.S. Congress. An FBI agent cannot flout the law." Cher asked about the separation of Church and State. Todd said, "Cher, the Supreme Court considered and rejected the argument, making the distinction that the Freedom of Religion Clause applies to religious beliefs and not overt actions that violate public policy."

Aside from the law, Cher asked how Elise felt about plural marriage. Elise responded, "Although I am Todd's only wife, should he marry, I would not be jealous. It's part of our faith, and he would not be cheating with another woman. We would be treated as equals and become friends." Cher understood the reasoning but said, "Your answer sounds logical, but how can your beliefs cure natural emotions of the heart?" After Todd's Fundamentalist explanation, the sizeable four-bedroom cooperative made more sense.

Todd glanced at his watch, realized the late hour, and said, "Why don't we call it for the evening. It's past midnight. We're tired and have dealt with quite a lot this evening. I'll be back from Washington in two days. While I'm away, James Ryan will oversee the Malik investigation." He handed her his card. "Cher, don't hesitate to call my cell number," and added, "We trust you'll keep our Mormon discussion to yourself. I'll update you when I return to the office."

Cher walked home and considered the juxtaposition of Elise and Todd's religion with blackmail. She shrewdly concluded that a criminal target could blackmail Todd because of his religious beliefs. She understood the vital importance of her pledge to secrecy and decided to filter what she told Mucky when answering his endless questions when she arrived at the apartment.

Chapter 5:

The Big Smell

Mucky was sound asleep when Cher arrived home. She celebrated her good fortune, made a bloody Mary, and sat on the couch with a sketchbook. Not giving much thought to it when she arrived home, Cher questioned why the hallway smelled like sour vinegar. After removing her shoes, she tip-toed to the hallway and found the odor coming from Malik's apartment. She returned to the couch and fell asleep with her open sketchbook.

While asleep, Mucky startled her in the morning when he jumped on the couch and sarcastically asked, "Did you and Todd try to solve every case in the City of New York last night? What time did you get home?" He chose the wrong morning to criticize. Cher scolded the old man. "Look, I'm tired of your attitude. I must make an important telephone call to the FBI."

She called Agent Ryan. "Mr. Ryan, this is Cher Egegian. Todd and I met last night, and I have some new information. Could you see me this afternoon? That works great; I'll see you at two." She returned to the couch, and Mucky asked, "What's the new 'information?" She told him she noticed a foul odor when she arrived home last night. Before Cher could continue, Mucky asserted his innocence. "Don't look at me!" Mucky remembered his first TSA security screening when he flew to his brother Saul's funeral in Longboat Key, Florida, and the embarrassing incident at the last checkpoint with his smelly socks. "Whatever you smelled didn't come from me. Don't report me to the FBI just because you're aggravated at me." Cher laughed and said, "Old Man, you're too defensive this morning. When I arrived home last night, there was a sour smell of vinegar in the hallway; I'm sure it came from Malik's apartment." Mucky said he didn't notice anything and, with a sarcastic edge in his voice, said, "Of course, I went to bed at a decent hour."

***

Cher allowed Mucky to attend the meeting at the FBI office with James, who greeted her with a smile and asked about the new information. Cher explained, "I arrived home late after the meeting with Todd and noticed a strong sour vinegar odor coming from Mr. Malik's apartment around midnight." James said, "The first thing that comes to mind is a meth lab. The drug produces a by-product with a similar odor absorbed by clothes and other fabrics. But only an idiot would produce meth in an apartment unless they had a death wish to deal with the DEA. I learned your neighbor has ties to Serbian organized crime. We know they provided the cash for his one-way ticket to New York; I'm still waiting for more State Department information. We may have sufficient probable cause to search his apartment and seize the hard drive from his computer. I'll pass on your information to Todd."

After they left the building, Cher told Mucky she recalled a similar smell as a child during a Sunday visit with her grandmother. Mucky said, "You mean to tell me your grandma was a meth dealer? Cher laughed and said, "Of course not, knucklehead. She liked to pickle cabbage and peppers."

***

When the two returned to the apartment, and while watching television, the landlady knocked on their door. Mrs. Schultz greeted Cher, but this was not a social visit. She came to warn them that Mr. Malik saw the tape of Cher in the parcel room the day she examined his packages. "He requested to see the videotape. Cher, you know tenants can make such a request if they are missing packages or suspect tampering with their delivery; I wasn't there. Paul Haldeman worked the front desk. He didn't accompany Mr. Malik to the video room as required. When Mr. Malik returned the key, he said, 'That busy-body girl in 803 was going through my packages.' That's all he said to Paul, who believed he looked overly excited and mad. Honey, stay away from that man, and please be careful." She thanked Mrs. Schultz and returned to the couch with Mucky.

Chapter 6:

Where Do We Hide

Mucky panicked. "I was right about that overgrown cow. Who knows what Malik did while alone in the video room because of the stupidity of the two-ton moron. We're probably on his most-wanted list. What do we do?"

Like a guardian Angel, Elise Montgomery happened to telephone to chat. After Cher told her what had occurred, she said, "Todd is in transit to New York. He will be home this evening. I insist you pack what you'll need for a few days and stay here until we can sort this out. Bring your little boy as well."

***

When Todd's government jet entered New York airspace, he called James Ryan and learned about the fiasco in the parcel room and the screw-up of allowing Malik alone in the videotape room. James said, "Boss, I went to the apartment. He didn't return the videotape. When the landlady and I went to his apartment, Malik had left and cleaned out the place. Forensics is there. When Cher called from your apartment, I assigned an agent to your home for protection." Todd said, "James, you did the right thing. Stay on top of it."

Todd expected to be home before seven. He told James, "See what you can do to arrange a nine o'clock conference call tonight with the Director of Homeland Security, the State Department, and the President's National Security Advisor. This will allow time for us to review what we have. Also, arrange a stenographer for the conference call."

***

Todd returned home to an overcrowded and noisy apartment and jested, "Elise, you shouldn't have thrown me a surprise welcome home party." His good humor didn't last after James briefed Todd in the office. The six-inch door slammed shut. "Boss, we missed him by thirty minutes. It's a shame because we just received the FISA search warrant. Forensics will be there all night.

"The CIA confirmed Malik's connection and funding from a Bosnia underground Muslim extremist group. We have credible intel confirming his parents were murdered when Serb forces attacked Muslims in Eastern Bosnia. The group blamed the United States for straddling the fence during the height of the ethnic cleansing of the Muslims from Bosnia. The CIA Station Chief in the area believed this caused Malik's recruitment and his radicalization by the underground group. They correctly claimed that the United States lifted the Bosnia trade embargo without consultation with NATO. I don't think we're dealing with a meth lab. We're looking at something more sinister."

Todd agreed, "During my closed-door testimony before the House Intelligence Committee, I questioned why the United States waited for over four years after the fall of Yugoslavia to take a leadership role to end the ethnic cleansing in Bosnia. We handed the terrorists ammunition to indoctrinate Muslims like Malik. He knows he's been discovered, but I believe Ms. Egegian is of little concern to our suspect. He has more immediate concerns than her." James agreed. "Sometimes I question who's watching the shop. I wonder how many others are in Malik's terrorism cell roaming Manhattan and if the operation was scrapped after Malik saw Cher on the video." As they were about to leave, the United States Embassy teletyped Jon Malik's United States Visa Application, his passport, and a recent photograph.

***

Todd and James rejoined everyone in the living room without changing their demeanor. The two men had great poker faces. The delicatessen across the street delivered dinner. Todd released the agent safeguarding Cher but told him to grab some food before he left. He explained that the case had taken a different turn and that she would be safe with him and Elise. The OMB had built a corridor from Todd's office to a separate secure residence with a kitchen, bath, and sleeping quarters, perfect accommodations for Cher.

***

The office telephone rang. Todd and James ran to the office. The forensic team at Malic's apartment found a metal canister containing an unknown chemical behind the refrigerator. The agent in charge of scouring the apartment told Todd the prints lifted from the cylinder did not match

Malik. Todd instructed the agent to have the Laboratory Director patch into the conference.

After the call, Todd told James, "If the owner of those prints has any booking record, we'll have his name and address within hours, and I bet Malic will be with him."

Chapter 7:

Government at Its Best

The critical video conference started as planned. Todd and James sat among the most powerful men in the world—The President's National Security Advisor and his Chief of Staff, the heads of the State and Defense Departments, the United States Attorney General, and the Director of Homeland Security.

Todd greeted everyone. "I thank you, gentleman, for gathering upon such short notice, but this matter is moving quickly. Everyone has been briefed and recently updated on all facets of this matter. Initially, our office believed the Jon Malik investigation involved the production of methamphetamine. I'm sure that's not the case. He's a political terrorist."

A notification flashed on the screen with a six-number code that allowed the New York Forensic Laboratory Director, Susan Bennet, to participate in the video call. Dr. Bennet, a Yale Ph.D., completed her analysis of the canister and the prints lifted from the container. The President's Chief of staff said, "Ms. Bennet, you have the floor. Please present your findings." She agreed with Todd. "This is not a DEA matter. We found a sealed plastic bag inside the canister containing a mix of nitroglycerin and sorbent pellets of cesium to release radiation and ordinary industrial resin for stabilization. I understand we know the subject received many boxes containing these materials. I project an explosive equivalence of 10,000 pounds of TNT, twice as much used in the 1995 Oklahoma City bombing. If detonated, unlike 1995, there would be radiation poisoning that would affect a large population near ground zero." Complete silence overtook all the participants. When nobody spoke, Dr. Bennet continued, "They would detonate the dirty bomb by cell phone, probably miles from ground zero."

The Homeland Security Director wanted to raise the security threat level. The President's National Security Advisor asked, "Director Montgomery, what's your opinion?" Todd disagreed. "I'm afraid raising the threat level contradicts what should be done. I always feared the subway system would be the next target. The Assistant Director and I discussed this before the call. When such scenarios like this were considered by the intelligence committees of Congress, they floated the idea of shutting down the entire subway system. Any government explanation for doing so would result in a catastrophic mass exodus from New York City. I opposed this because we have no actionable intelligence that the subway system is a target. We need to buy some time until we receive the National Fingerprint Identification System results and catch these people. Dr. Bennet believes the trigger for such a dirty bomb is a cell phone, and I agree. My suggestion: shut down all the cell towers. Only 9-1-1 calls could be made. We explain the shutdown as a preemptive measure concerning an investigation of a computer hacker trying to compromise the cellular system. A catastrophic panic is eliminated and buys us the time to find these people."

The President's Director of National Security asked Todd, "Agent Montgomery, how much time do we need to buy?" Todd looked at James Ryan and winked before his answer. "We have the known prints of Malik and the prints of a probable accomplice. If law enforcement ever booked this 'John Doe,' we'll have his identity within hours. Malik's photograph has been distributed to all the metropolitan airports, Grand Central Terminal, Penn Station, rental car agencies, and all toll road stations in the metro area. I'd bet good money that Malik is with the owner of the second set of prints. If we find him, we find Malik. We're talking hours and not days." The conference recessed with all in agreement with the cell tower shutdown. The Attorney General was asked to check the legality of the government's unilateral shutdown and potential exposure to civil liability.

Chapter 8:
The Well-Crafted Explanation

Todd spoke to Cher in his office. "We're almost certain Mr. Malik is not in this country to make and sell drugs. He vacated the apartment, but in his rush to leave, he left evidence pointing to his participation in a political terrorist attack. We found prints of another individual, and we'll know his identity and location shortly. Cher, you're not in any present danger. Malik is more worried about our Agency than you. I released the agent assigned to you, but you shouldn't return to your apartment until we check our leads. I'd feel more comfortable if you and Mucky remained here for the next forty-eight hours. Elise would love your company." Cher agreed but asked, "What does Malik plan to do?"

Todd hesitated with his answer. "We're not sure. However, past acts of terrorism have involved using cell phones as detonators. We all agreed that all cell services should be shut down. Except for the police, firefighters, and nine-eleven emergencies, you'll need a landline to make telephone calls. I'm sorry, but I can't discuss any more detail."

Cher smiled and said, "Without my cell phone—I'll feel lost. May I pour a glass of wine before bed? Mucky is already asleep." Todd said, "No need to ask. Elise went to bed as well. I'm going to work on my notes from the evening conference. Whenever you want to call it a night, I'll show you the separate entrance to the secure living area."

When Todd finished his conference notes, he walked into the kitchen and saw Cher working on dress designs in the living room. "I'm surprised you're still up. Do you realize it's two in the morning? I didn't know you were a night owl." Cher sounded tipsy to Todd when she said, "No, I just can't sleep. The quick pace of the day has me worked up."

Cherie fell asleep on the couch. The following morning, Mr. Muckelman again provided his identical and cynical wake-up service, jumping on her and asking, "Are you boycotting your bed? What happened last night? I fell asleep during the conference call." After Cher relayed Todd's explanation, Mucky said, "I just can't seem to catch a break in this world. First, my airplane nose dives to the ground. Instead of dying, I suffer the indignity of having four legs and a tail, nearly drowning in the Atlantic during our boat trip from Longboat to New York, and now you tell me this place will blow! Let's get out of town!" Mucky jumped on Cher's lap and continued, "At least we should warn George and Maria."

Cher forgot to ask Todd about the foul odor she had noticed in the hallway. When she mentioned her lapse to Mucky, he said, "Quit trying to pin that on me! How many times do I have to repeat myself?"

Chapter 9:

Running from the Pepper & Pickle Man

Cher caught Todd before he left for work. She wanted to visit George Markarian and explained, "He's a close friend, a successful businessman, the Executive Director of The Muckelman Fine Arts Foundation, and the source of my grant to the Fashion Design Institute. They live in the Trump Tower." Todd said if Cher felt safe, he would have no problem with the visit, but discussing Malik would be off-limits." Cher agreed.

When told about their visit, Mucky was so excited he started running in circles, but the old man had to stop. Cher said, "See what all that cigarette smoking did. George is having us picked up at noon. Todd said there couldn't be any discussion of Malik. You can come if you promise to keep your mouth shut." Mucky agreed.

Maria had set the patio table for lunch. The first words out of George's mouth stunned Cher. "Big news—did you guys hear about the terror situation? They've shut down all cellular systems to prevent a cell phone detonation of a dirty bomb and issued a phony reason for it." Wanting to honor what was now a moot pledge to Todd, Cher said, "What? You're kidding. There's nothing in the news. Fill me in." Mucky whispered to Cher, *You're a sharp cookie to have George do all the talking.* George gave Cher his disarming smile and said, "Cher, there won't be any news either. The government is intentionally misleading the public to avoid a panic bolt from the Borough. You know, George Markarian is always the first to know." He continued, "There is some foreign national, and Mucky, for your sake, we'll call him a foreign element with a radioactive bomb rigged somewhere in Manhattan." It was as if the shrewd Armenian participated in yesterday's secret conference.

Mucky wanted to head for the hills. "Let's get out of this hell hole! Let's go to Tokyo!" George laughed at the statement. "Old buddy, that's not a good idea. We would stick out like a sore thumb, and you could be grabbed for someone's dinner. Japan is a bad idea. Any industrial capitalist country is a target, and Tokyo is nothing more than Manhattan on bad steroids."

Cher changed the subject, fearing an accidental linkage to Malic and the FBI investigation. "George, I noticed a smell at the apartment a few days ago." Mucky interrupted. "George and Maria, for the record, it didn't come from me." Cher told Mucky to hush up, and she continued, "It was a strong sour smell close to vinegar, but much stronger. Someone mentioned a meth lab produces a similar residual odor, but that has been ruled out. Any idea what it could have been? I told Mucky it reminded me of my childhood memory of visiting my grandmother when she pickled peppers and cabbage. The smell never returned." George laughed and said, "I think you answered your own question. Eastern Europeans, my family included, loved to pickle—nothing to worry about."

Everyone had an enjoyable afternoon while Mucky criticized the money changers sixty-four floors above Wall

Street like he had done with dentists, calling them 'money-hungry scoundrels' who lived on the fancy East Side of Long Island.

Chapter 10:

Operation Overlord

The FBI Laboratory identified the owner of the second print from a recent Queens County booking for public disturbance and a felon in possession of a handgun. Ironically, the suspect, Ahmed Hudzic, lived at Coolidge and Eighty-Second Drive, eight minutes from the Queens County Criminal Court. The area looked like the opening shot of the popular television show The *King of Queens.*

When arrested, the police found a 9mm handgun holstered to Hudzic's ankle. With his lawyer able to reduce the previous bond, he left the courthouse after his arraignment. Hudzic lived at the same address the FBI had targeted for twenty-four-hour surveillance after they lifted and identified his prints and location from the FBI database. Todd's prescient recommendation made at the conference proved to be correct.

Todd, James Ryan, and the United States Attorney met at a coffee shop in Queens. After a brief chat, they walked across the street to the courthouse for their appointment to discuss Ahmed Hudzic with the Queens County District Attorney. They discussed the terrorist plot and the involvement of Mr. Hudzic.

The New York District Attorney agreed with the bond revocation. He said, "I argued against the lower bond during the State arraignment. We'll have a different Justice this time." The U.S. Attorney brought certified copies of the government's documents, including the Information that charged the defendants with conspiracy and multiple terrorism counts. Todd said, "We're ready to present the Information to the Grand Jury for a formal Indictment." The DA was on board. He notified Hudzic's Public Defender and scheduled an expedited chamber hearing that morning.

The U.S. Attorney, through the DA, offered the Court a certification of the FISA warrant, the FBI Laboratory's Director's affidavit establishing a match of Hudzic's prints to those lifted from the canister, the Federal Information, and the Homeland Security Director's affidavit request to seal the file.

Mr. Hudzic's public defender silently sat in Court. The Judge addressed him, "Mr. Vanderfield, Why is your client not in the courtroom?" He told the Court his office assistant saw him in the building. "He's probably lost." The Judge said, "That alone is sufficient for a bond revocation." Satisfied with the evidence presented, He ordered the bond revocation, granted the relief requested, and issued an arrest warrant.

***

Todd and James returned to the office, which buzzed with activity and resembled the intensity of General Eisenhower's war room during World War Two's D-Day, crossing the English Channel to disengage Germany's occupation of France.

The agents reported continued activity in the house, confirmed by the burning evening lights. No one had entered or left since the start of the surveillance. Todd disclosed the agenda to the lead Agents. "We will seal off two miles of the perimeter at 1:00 a.m. this morning. The lead agents will report at their assigned command posts at 2:00 a.m., and the Special Weapons and Tactics Team will approach the residence at 3:00 a.m. The Helicopters with special forces will arrive when the swat team is at the door. The Attorney General will notify local law enforcement two minutes before the assault. The President will be in the war room during the entire operation.

"We all know about the radiation fallout that results from a dirty bomb. Our intel strongly suggests these are not suicide bombers; if they were, they would want a detonation to occur with the most impact, and that's not a small house in Queens. However, properly suited bomb technicians will enter after agents have neutralized the suspects. Your orders are to place restraints on all those inside and, in your judgment, use all necessary force you deem necessary. Surveillance has counted ten to twelve suspects, with no children or women in the house. We may have surprised an entire terrorism cell. Go get them. Godspeed and good luck."

The lead team lifted the Enforcer, a 16 Kg battering ram of hardened steel capable of ramming a locked door with three tons of impact. After surrounding the property, an agent radioed the code, "Go Charlie One," and the FBI paramilitary forces advanced on the ground. The military copters swarmed in the air with search spotlights bright enough to light a baseball stadium, the Enforcer obliterated the front door, and agents entered with MP5 submachine guns. The occupants offered no resistance. The FBI completely surprised the suspects. Among those chained and taken to federal detention were Jon Malik and Ahmed Hudzic. Horrific terrorism, as deadly as nine-eleven, had been averted by a girl in fashion school and a Garment District talking dog.

Chapter 11:

The Toast of the Town

Maria and Mucky spent the night at Trump Tower. They learned about the capture of the terrorist cell while watching the Today Show. The producers of the morning telecast chose Lester Holt, NBC's national and local anchor for New York City's newscast, as their host for the evening national telecast to report the successful capture of those involved in the lightning raid in Queens.

Holt credited a New York City college student and her dog with notifying the FBI. The report surprised Cher, and Mucky became upset because their names were omitted. George looked askance at Cher and said, "Well, yesterday, you were pumping me for information to determine how much I knew. Shame on you." He next looked at Mucky and warned, "Be careful about what you wish. If the news media learns your identities, and *The New York Post* probably knows by now, you can kiss your privacy goodbye."

Cher apologized to George and Maria for not telling the whole story, saying, "A Serbian national, Jon Malik, moved into the apartment across the hall from us. He acted peculiar when Mucky and I baked a plate of cookies to welcome him to the area." Mucky said, "I told Gucci girl about the foreign element's chemical deliveries, and she snitched to the FBI Director we met at Mr. Biggs. This weirdo saw Cher examining and taking pictures of his parcels in the surveillance tape. Mistakenly left alone in the taping room, he stole the video and emptied his apartment. An FBI agent took us to Todd Montgomery's home with a safe house within the living quarters."

George interrupted Mucky. " Muck, The only thing missing from your narrative is the blond girl who had dinner at the Bronx Rib House with your married friend."

Cher continued. "Without a cell phone, cabin fever struck, and I wanted to see you. Todd allowed us to see Maria and you but made me promise not to mention anything about the investigation. What was I supposed to do? How did you know about the terrorists' plans?" With a sardonic smile, George said, "Someday, I may tell you."

***

When Mucky and Cher left for home, George walked them to an awaiting limousine. In the lobby, Donald Trump watched all the press activity and saw Cher and Mucky. He blurted, "I bet you and that dog are the heroes that saved New York." Reporters overheard Trump's comment, and a frenzy of newsmen overwhelmed the Tower's security. With a tight hold of Mucky, George, Donald, and Cher ran to Trump's waiting private elevator. He insisted they use his helicopter on the Tower roof and said, "This is all my fault. I'll call my pilot; he'll be here in ten minutes."

Chapter 12:

The Grand Arrival

The sound of a helicopter landing on the roof awoke Todd and Elise. Todd said, "I've had it with choppers. What a wake-up call." A knock at the door followed. Cher, holding Mucky, smiled and said, "Todd, Good morning, and congratulations; you were great." Cher realized the pre-dawn raid had left little time for Todd to sleep and apologized, "I'm sorry we woke you. I'm so proud and relieved you guys at the Bureau avoided a catastrophe. We watched the report on the Today Show."

Pleased with her safe return, Todd said. "I'm never too tired for you. Who knows what would have happened if it wasn't for your vigilance? Was that you landing on the roof?" Cher said they couldn't get through the throng of reporters, and Donald Trump offered us the use of his chopper and pilot.

Elise grabbed her robe and joined everyone for coffee. Cher told them that George knew about the terrorist threat and swore, "There was not a peep from me. Five minutes after we arrived, the first words out of George's mouth involved the terror threat. I let him do all the talking.

"You've got to meet George and Maria." Todd also wanted to meet them and asked Cher to arrange a convenient evening for dinner. "I've heard so much about them; I'm quite excited to meet them." His reason was less social and had more to do with how George knew about the secret operation.

****

During breakfast, Cher received a surprise suggestion from Todd that stunned her. "Why don't Mucky and you move in with us? You'll be safer here than in your apartment." Elise suggested, "We can tour the building and later pick your bedroom." Cher said she would consider the offer, and Mucky experienced a rare speechless moment.

****

The building tour with Elise impressed Cher. She said, "Your place is so nice. The conference rooms and lounge in the lobby are impressive. You have a doorman, a front desk clerk, and a parking garage." Mucky couldn't resist repeating a sarcastic remark about the gentrified area. *Gucci girl, I was right. These new joints all have a sleeping front desk clerk, and I bet when we ask the doorman to hail us a taxi, he will have his hand out for a tip faster than Wyatt Earp could draw his six-shooter. We'll have to wait until it rains for the leaking garage.*

I said a studio would cost four grand a month. Todd and Elise have four bedrooms on the top floor. They could put a kid through Yale for what they're paying.

Unable to suppress her amusement from Mucky's comical comment, Elise noticed Cher's unexplainable reaction to

Mucky's inaudible comment. Caught flat-footed, Cher offered a lame explanation. "I chuckled because the pool reminded me of a friend's story, where a klutz stood in a Central Park rowboat and fell into the water. When he asked for his money back, the attendant mocked the drippping man, refused a refund, and said, only fools and George Washington stand in rowboats."

***

Cher thanked Elise for the tour and said she would seriously consider her kind offer. Cher and Mucky left for her clinical work-study at Cushman's Clothing. The doorman hailed them a cab, and Mucky said, "I sure was right. Did you see the speed of *Quick Draw McGraw* with his outstretched hand?"

Chapter 13:

The Proud Boss

Although late for work, the boss, Bernie Cushman, greeted Cher like a celebrity. She asked for a few minutes in his office for an opinion. Bernie said, "After all the great advice you gave me about the shop and how you saved the City, anything you want." Mucky said, *"Ask him for a raise and see if he's a man of his word."*

Cher explained to Bernie how she became involved and said, "One of the terrorists recently moved across the hall from my apartment. We already knew the Director of the New York FBI. Mucky and I met him and his wife at Mr. Biggs restaurant and bar. Mucky and the apartment manager drew my attention to the new neighbor's unusual behavior. During an FBI interview, I related all I knew about the man, and the Director placed Mucky and me in protective custody. I stayed in a safe room at his huge four-bedroom apartment.

"We became close friends, and his wife asked me to move into their apartment this morning. Once I graduate, my grant money ends, and moving into their home eliminates the worry about finding a job and making a budget for new living costs. What do you think?"

Bernie did not take long to share his thoughts. "The character of your friends is not an issue. The decision really comes down to how you feel about the arrangement. If you are comfortable with these folks, your decision is easy. Accept the offer, but as I always say, have a plan B. People can change over time. You or your friends may tire of each other." She thanked Bernie and walked downstairs to draft some new designs. Mucky looked at Cher and said, "I guess my opinion doesn't matter." Cher soothed Mucky's bruised ego and said, "Your feelings matter the most. I had planned to discuss the matter with you after I spoke with George because the Foundation pays the rent."

Chapter 14:

The Second High-Level Conference

Wanting a quick resolution of where to live, Cher called George, who agreed to stop by her apartment after work. Although skeptical about the move, Mucky didn't object because he trusted George's judgment. He said, "They're good folks, and their home sure beats our present place. I wish you would tell them about me." Cher refused to discuss the subject and avoided Mucky's pestering insistence by not foreclosing the issue. "Let's solve one problem at a time." Keeping the issue open satisfied him for the moment.

When George arrived at Cher's apartment, he asked if she enjoyed the copter ride. She said, "A bit scary, but I worried Mr. Trump might join us. He sure likes being in the limelight." George agreed but defended him. "Everyone has their faults. So, what do we need to talk about?"

Cher said, "Todd and Elise invited Mucky and me to stay with them. They live in a new four-bedroom building at Forty-Fifth Street and Eleventh Avenue. I'm considering their offer, but I know the Foundation pays the rent, and I wanted your input."

George didn't see any problems with the Foundation. But he questioned whether Cher considered her privacy and how Mucky would react. "I have no problem with the arrangement as long as you consider those issues, and keep your apartment as a backup. Events sometimes disappoint." Cher smiled and thought: *Pretty close to what Bernie said.* Mucky added his two cents, "George, that's exactly what I thought."

Cher almost overlooked the dinner invitation. "George, I appreciate your time and sound advice. Give my love to Maria. Wait, I almost forgot. Todd and Elise have heard me brag about you guys and asked me to invite you to dinner at their home." George smiled and said, "Sure. I never turn down a free meal. I'll talk to Maria tonight and call you. George looked at Mucky and asked, "Old buddy, I sure hope you're behaving?"

***

Cher told Elise about the confirmed dinner plans with George and Maria and said, "Elise, I would love to live with you guys. My only company is Mucky, and he's not much of a conversationalist."

When Mucky entered the living room, Cher carried him into the bedroom to tell him about her decision. "Well, Bernie and George advised us to move if we wanted. I still don't plan to tell Todd and Elise about you, at least not until we are well settled with the move. I'm going to keep the apartment as a backup. One never knows. George said the Foundation would continue to pay the expenses until graduation." Mucky liked the plan. "I feel like one of those East Long Islanders. We have two homes—a fashionable top-floor Manhattan apartment with a sleeping front desk clerk and our doorman buddy, and when in the mood, we can slum at my old place." What bothered Cher was the planned move sounded too good to be true. Mucky disagreed. "That old saying is out of date. Like the one 'If it ain't broke, don't fix it.' Lifetime New Yorkers believe—if it's too good to be true, you better grab it before someone else does, and if it ain't broke, break it—something new is better than something old." The old man's contorted logic remained consistent.

Chapter 15:

Guess Who's Coming to Dinner

Cher telephoned George concerning their dinner invitation and told him she had not revealed anything to Todd and Elise about Mucky's transformation. She said, "I know we occasionally slip and forget that only Maria, you, and I can hear the old rascal. His description of things and events is so unusual they're funny. I laughed at something he said the other day during Elise's building tour and received a puzzled look about my behavior. We have to either remember to control our reaction, bite our lip, change the subject, or walk away to stretch our legs." George asked if Cher ever planned on telling them, and she replied, "George, I'm not a multi-tasker like you. I take one thing at a time."

***

The morning of the dinner party, George telephoned Cher and offered to drive Mucky and her to Todd's, but she declined and explained, "I'm leaving early to help Elise prepare." George said, "Call when you're ready to leave, and I will send a driver. It's pouring outside; you don't want to walk in this nasty weather." Cher accepted George's kind offer.

***

The heavy rain continued when George's driver arrived at Cher's apartment. The chauffeur drove to the garage walkway of Elise and Todd's building. While in the elevator, Mucky smiled and said, "Gucci girl, I was right! I hit the trifecta! We have a leaky garage!"

Mucky hopped on the couch and fell asleep while the two women worked in the kitchen. A prime rib roast crackled in the oven, and the aroma woke Mucky, who sauntered into the kitchen for a peek. Cher peeled potatoes, and Elise prepped the asparagus tips and hollandaise sauce. Elise said, "All that's left to do is to mash the potatoes and prepare the gravy and salad. I've already set the table. Let's sit in the living room and relax."

Elise handed Cher a Merlot, laughed, and said, "Mucky sure knows how to relax. He's again asleep on the sofa." Cher remarked, "I think we'll all have a great time together. But tell me if you or Todd later have a change of heart because I can keep my apartment until I graduate from the Institute."

Mucky heard his name and jumped on Elise's lap when she asked, "What kind of dog food does Mucky like?" He stared at Cher and prayed: *Please give her the correct answer.* Cher pleased the old man. "He doesn't eat dog food. He eats whatever I'm having. I cut everything up in a bowl, and we walk after each meal." *Good job, Gucci girl. Her cooking smells great.*

***

Elise joined Cher with a glass of wine and said, "Before you comment about Mormons and alcohol, Todd and I occasionally partake of wine and spirits." Mucky liked what he heard. *Elise, you and I are going to get along fine. I do the same thing with Jewish dietary rules to keep up appearances of being a good Jew. Gucci girl, we have to talk later about my occasional nip of alcohol.* The evening looked promising to the ornery dog.

Chapter 16:

The G-Man and the Billionaire

When George and Maria arrived for dinner, Mucky ran to them. George picked up the little guy and whispered, "Hey, old buddy, how are you?" Mr. Muckelman warned George, *"Cher hasn't told them about me yet. Don't let them know you can hear me!"* As he set Mucky on the floor, George winked his acknowledgment. He next walked to Todd, placed both hands on Todd's shoulders, and said, "Cher, I finally get to meet the G-Man you've been raving about." He next turned and admired Elise, Maria, and Cher and asked, "Aren't we blessed to be in the company of such beautiful women this evening?

"Cher told me about your successful capture of the terror cell. Great job! Cher continued questioning how I knew about it, but I bet you're more curious than her. I believe you're familiar with your predecessor, Bill Hart. Bill and I have known each other for years. He's the new Chief of Police of the eastern Long Island Village of Sagg Harbor. I was at his office checking on someone's background and learned about the plot from him." Relieved that he did not have to raise the question, Todd offered wine to everyone and said, "George, I was a bit concerned about the breach. I've met Bill. He was a good New York FBI Director. As they say, "No harm, no foul, because the entire terror cell is in custody." Todd pointed to Cher and continued, "We have that young lady to thank. Let's toast our good fortune for Cher's diligent behavior." Mucky expressed his displeasure. *Gucci girl, I'm the one who told you about that terrorist. I feel like a fire hydrant that's become an overused relief station for all the dogs of Manhattan.*

***

George and Todd shared an interest in exotic cars. When they were about to leave for the garage, Elise warned, "Unless you men like your meat well done, I suggest you play with your toys later."

The dinner conversation shifted to Cher and her decision to live with Todd and Elise. George said, "I sure hope your closets can accommodate our favorite compulsive shopper." Mucky looked up from his dinner bowl and agreed with George, *"Ain't that the solemn truth."*

Everyone had finished dinner except Mucky, who demonstrated his relentless effort to remove every morsel of meat from his rib bone. Everyone else took their coffee and dessert in the living room.

***

After their obligatory stay with the women, the men left for the garage. George's Mercedes impressed Todd, who drove George's red speedster to his reserved garage location. After the brief test drive, Todd said. "Your car is a racing beauty." George gave Todd the racing hat that flew off Mr. Muckelman's head while driving from the airport. Todd asked, "What speed have you hit with this monster?" George said, "Two-hundred and five at the track, but you can't sustain that speed."

 When George saw Todd's car, he asked, "What year is your Roll's?" Todd answered, "I bought the car a year ago. Before you start jabbing at me, I'm not on the take, my wife isn't loaded, and they haven't raised the salaries of FBI Directors. I made my money the easy way, a large inheritance from my uncle."

George looked inside the new Rolls Royce, saw a portable oscillating light and siren switch, and asked, "I see you use the Rolls at work. I bet you get some looks. Have you ever been pulled over for speeding?" Todd said, "With all the traffic gridlock in the City, it's impossible to speed in Manhattan. I've been stopped on the highway when not responding to a call. However, the traffic officers either knew me or gave me a pass after seeing my credentials. Occasionally, I receive odd looks from the officers, trying to reconcile how an FBI Director can afford to cruise town in a Rolls Royce." George laughed and said, "I bet you do, old buddy."

***

When George and Maria returned home, Maria said she noticed how Todd looked at Cher during dinner. George had no clue what Maria meant. "I don't understand what you mean by a look. Guys don't notice those things. Why don't you just tell me what's bothering you." Maria replied, "Just forget it. I can be too critical at times. I probably misinterpreted what I heard and saw."

Chapter 17:
Life with the Montgomery's

Days after the dinner party with George and Maria, Cher noticed a change in Mr. Muckelman's behavior. When alone with him, she asked, "You appear troubled. Is there something bothering you?" Mucky remained quiet, and Cher continued, "I can't read your mind, thank God. You have to tell me what's going on." Mucky aired his complaint. "I am still bothered by your refusal to tell them about my transformation. I feel more like a dog every day. I didn't receive any credit for tipping you off about Malic. What are we going to do about it?" Elise walked into the room, and their discussion ended.

Cher understood Mucky's feeling of isolation and realized she had to include him: *I wish I knew what to do, but I cannot go through the rigor of how we convinced George and Maria.*

While Cher prepared to leave for Cushman's work-study, Mucky entered the bedroom. "Gucci girl, would you mind if I didn't go to the store with you today?" Mucky had never missed a day of work with Cher. He continued, "Would you see if George or Maria can pick me up for the day?" She reluctantly agreed and knew she had a severe problem.

***

Maria arrived to take Mucky to the Trump Tower. During the drive, Mucky shared his feelings about their new living arrangement with Todd and Elise. "I've never seen Elise mad nor raise her voice in anger. She's like one of those 1950s television moms. The worst part of living there is that Cher and I cannot freely talk to each other. She's afraid to tell them about me. She doesn't want to sound like a crazy person." Maria understood Mucky's feeling of isolation but said, "Don't forget, in the beginning, George and I thought Cher was either nuts or a grifter. You know she doesn't like this either." Mucky agreed and said, "That means we're both miserable." They continued their discussion with a terrace lunch on a warm, sunny September day. Sitting on the tower's top floor, Mucky felt like the overseer of New York City.

George arrived home at seven and saw Mucky in the living room with Maria. He smiled at Mucky and asked, "Old Buddy, where's Cher?" Maria explained Mr. Muckelman's problem. George realized he had never seen Mucky without Cher nearby and said, "This is a first." He asked Mucky, "Have you arrived at any solution?" With Mucky's glum appearance, he had not. "George, I hoped you'd come up with something," Mucky asked to spend the night at the Tower. "Would you call Gucci girl so she doesn't worry?"

Chapter 18:

Trouble on the Horizon

When Cher returned from work, Elise noticed her forlorn appearance and questioned, "Are you feeling all right, and where's the little boy?" Cher said, "I'm just tired." She explained Mucky's absence. "He missed Maria and George and wanted to spend the night." Unlike her reaction to Mucky's comment at the swimming pool, Elise masked her surprise at Cher's ability to know what the dog wanted.

After dinner, Cher left the co-op for Mucky. During her absence, Elise questioned Todd about Cher's unusual behavior with the dog, and Todd asked, "Where is Mucky anyway?" Elise said, "She's picking him up from the Tower. I've wondered if you've noticed anything odd about Cher and Mucky?"

Todd smiled and said, "I've noticed Cher's outstanding beauty and sexy olive complexion. I love coming home to two beautiful and intelligent women. I wish they would become part of our family, but that's it." Elise continued, "This is more about Mucky. During the building tour, I noticed she appeared to react to Mucky as if he had said something to her, and tonight, when she arrived home without him, she told me he wanted to have a sleepover last evening at Trump Tower. How could Maria have known what he wanted?"

Todd recalled the terrorist ordeal and his interview with Cher, saying, "She had just introduced herself to Malic. I asked if she saw anything of interest in his apartment, and she told me Mucky walked into the apartment unnoticed and saw a computer. When I questioned how Mucky conveyed his observation, she claimed nervousness, and her answer seemed reasonable. I see your point. What do you propose we do?" Elise offered a flippant reply. "You're the FBI Agent; what do you suggest we do?" Todd said, "Real cute, but whatever we decide, we should talk to her together."

***

When Cher arrived at the Tower, George and Mucky greeted her at the door. Her cheeks were red from the brisk October walk from the West Side of Midtown to the Upper West Side of the Trump Tower. Mucky said, "Gucci girl, George, and Maria are returning to Longboat Key for the winter."

Maria joined them in the living room and said, "I heard Mucky deliver the news. We'll probably stay for the winter. The chilly wind off the Hudson always reminds me it's time to leave. Why don't Mucky and you join us?" Cher fancied the offer but declined. "I'd love to join you guys, but I have school and work-study at Bernie's shop." George discounted her worry. "If I call the school and Bernie, you can take a few weeks off." Cher enjoyed her visits to the Florida Gulf and fell in love with George's beach-front home but hesitated and said, "I'll sleep on it and decide tomorrow."

After dinner, George, Maria, and Cher enjoyed some aged cognac, and Mucky had his usual gin and tonic. When the chimes on the grandfather clock sounded, Cher looked at her watch and said, "Where did the time go? It's eleven, and I have an early day at the shop." Maria called a driver and insisted that Cher not refuse the offer. "Not a peep from you. The temperature has dropped, and it's too cold to walk home. Please consider Florida. We plan to leave in a couple of weeks."

Tired and not in the mood for a Spanish Inquisition from Todd and Elise, Cher was grateful they were asleep when Mucky and she arrived home. She picked up Mucky and tiptoed to their bedroom. Just before shutting the lights, Mucky asked Cher if they were going to Longboat Key. She nodded yes with a caveat: only if George clears everything with the school and Bernie. Elated with Cher's decision, Mucky wanted to hug her, but a long lick on her face sufficed.

***

When Elise saw Mucky the following morning. She fawned over him, the part of the living arrangement he loved. Todd had left for work before dawn. During breakfast, Elise said, "Isn't it wonderful to have little Mucky back?" The answer from Cher saddened Elise. "Last night, George and Maria told us they were leaving for their winter home in Longboat Key and invited us to join them. We won't be gone long." Elise feigned a frown and replied, "We'll miss you both, but I think you deserve some R&R after everything you've been through."

Todd learned of Cher and Mucky's tentative travel plans when he returned home for dinner. This interfered with his timetable to convince Cher to become part of the family. He questioned: *What can I do in two weeks.? Why do I feel Cher is saying goodbye forever?* Elise understood her husband's preoccupied and distant manner. He complimented Elise for a great dinner and excused himself to work in the office. Cher failed to notice his reaction to the news.

While reading in bed or, more accurately, staring at a book, Todd's sullen mood had not improved. Elise told him to ease up. "You're making too much of their trip, and it may become counter-productive if Cher picks up how you feel. There is no hurry to rush things." Todd kissed Elise and said, "Good night. I hope you're right."

In the morning, George called and told Cher that neither the Design Institute nor Bernie objected to her trip. Cher would be returning to Longboat Key, Florida.

Chapter 19:

The Big Decision

Cher correctly anticipated her first day on the Florida West Coast, which had eliminated the pressures of Manhattan, the foreign terrorism plot, and the difficult decision to live with Todd and Elise. Dinner at a seaside restaurant also helped the cause, as did George and Maria's ability to converse and control the uncompromising four-legged tailor from the New York Garment District.

The beach house had not changed, nor had the couple's early morning walks along the beach or the late afternoon poolside happy hours next to the waterfall.

All of the shop and restaurant owners of Longboat Key knew and respected George, but not necessarily because of his sizable wealth. He had either helped or offered financial assistance to many local merchants. When George entered any Longboat establishment, they treated him like royalty. He never waited for a dinner table, regardless of the day or time.

***

While Maria and George left to stock the bar with liquor and food, Mucky and Cher conversed privately about their problem. "Mucky, how are we going to resolve our living situation? I know you're unhappy, and I'm having second thoughts myself." Cher further endeared herself to Mucky. "Gucci Girl, you have proven yourself to be a considerate person. Few people would consider my feelings. My big problem is, don't you think we've lost the freedom we had before? We can barely talk to each other anymore. Don't misunderstand. Todd and Elise are fine folks, but it's not how it used to be."

Cher believed the time had come to tell Mucky about Todd's proposal, omitting any reference to polygamy, to honor her earlier pledge.

"Todd has asked me to become part of the family." Cher's explanation failed to register, and Mucky asked, "You mean they want to adopt you? There sure is a lot of adoption talk these days. What would your parents say?" Cher had to be more direct with the obtuse Pom. "Of course not. Todd wants to have a romantic relationship with me." Mucky looked like he had run into a brick wall. "But what about Elise? You can't do that and become one of those homewreckers." Cher called the relationship an open marriage and continued her pledge to Todd and avoided calling the arrangement a polygamist marriage. The old haberdasher remained flabbergasted.

***

George returned home after a horrible round of golf, but he had exciting news for Cher. "Maria, the damned sport should be banished. I was the worst of the foursome, fourteen over par." Cher and Mucky joined them in the living room. "Everyone, please don't ask about my golf score today.

"The big news is that Henri Durand of the House of Dior played in the foursome and wants to talk to Cher. He's the assistant to the fashion house's principal designer, and he said they're introducing a new traditional line of women's clothing. Your name came up when I mentioned your fashion portfolio and The Design Institute work study in the Garment District. Evidently, the school has quite a reputation. He returns to Paris in a few days and wants to talk to you before leaving." Cher was so excited she could barely put two sentences together. She erupted with her answer. "Oh my God! Thank you, George! I must be dreaming!" George handed Henri's hotel and telephone number to Cher and said, "Good luck, kiddo."

***

After Cher's persistence, George drove Cher to the Longboat Key Club Resort. As they pulled into the circle drive, the valet handed George his parking ticket, and Cher said, "Henri's hotel isn't too shabby." After George introduced Cher, he said he would wait at the lobby bar. Before leaving, he warned Henri, "You are not permitted to discuss my golf game."

Henri wanted to know the reason Cher chose fashion design. She said a chance meeting factored into her decision. "Of all places, I was killing time looking through a window of a tattoo parlor. An elderly man introduced himself, and his first words were, 'You'll never see a model with a tattoo.' He said I had the appearance and bone structure to become a Runway Model. I had no desire for a tattoo. Irving L. Muckelman, a man in his seventies, worked in New York's

Fashion District his entire life. I could tell he loved his work."

She said their decision to have a simple cup of coffee lasted for hours. "Mr. Muckelman described New York Fashion Week as a peek at the greatest trends in clothing design and a magnificent festival filled with endless energy. I think that hooked me."

Henri asked, "Why the portfolios if you were not in fashion design." Cher smiled and said, "It's my hobby. I will make the gown if I create an interesting and different design." Henri liked her answer and loved her ability more after looking at her sketches. They left to find George.

***

Henri had a busy schedule for the day, but he had sufficient time to join George at the bar for a glass of wine. Henri used the time to implore Cher and George to return to Paris and meet the design staff. The fashion designer felt George's presence in Paris increased the odds of Cher saying yes. George said, "Cher, you can't pass up this opportunity of a lifetime. Maria can watch Mucky. I haven't been to the City by the Seine in years. I'll stay out of the way because I'm aware of the temperamental behavior of you fashionistas." Henri laughed and joked, "George, assessing your behavior during the golf foursome, I've concluded the mercurial trait is likewise shared by business tycoons." George had the last word. "Cher, you'll be shaking many hands at Dior. Make sure you count your fingers before leaving."

George asked Henri to stylize an evening chiffon gown for Maria and gave him her measurements. He said, "Henri, old buddy, we made significant progress today. When Maria and I return to Paris, she'll be surprised with her Dior gown, and The House of Dior will have snared Cher for an evaluation by your design group."

***

On the Florida Gulf Coast, Paris predominated the conversation during dinner at the Longboat Key Dockside Restaurant. With perfect weather, Mucky joined everyone for dinner alfresco. He wanted to stay in Florida while George and Cher were in Paris. "I have enough trouble staying at Todd and Elise's with Cher, but it would be too strange to be there without her. I'd also be here to protect Maria." Maria resisted laughing, but George couldn't refrain from remarking, "You would make a perfect watchdog. If someone broke into the house, you would carefully watch the burglar's every move while hiding." Mucky replied, "Everything is a joke to you." At the end of the evening, Maria and George agreed Mucky should remain in Longboat Key.

Chapter 20:

Kismet

One week after her eventful sojourn ended in Longboat Key, Cher arrived at the Montgomery Home without Mucky. Todd and Elise speculated that her news would not be good. Gucci Girl and the possibly newly minted Dior Girl said, "I'm going to Paris for an interview at the House of Dior. I'm not sure when I'll be returning. This all happened because of a golf game. The assistant designer at Dior was part of George's golf foursome and saw my dress designs— kismet at its best. I'll miss you."

After relating the fantasy-like story, Cher rushed to Cushman's clothing ahead of Henri Durand to tell Bernie the news. Bernie had mixed feelings but understood the profound opportunity for Cher. She asked Bernie's opinion about leaving school early. He dismissed her concern and advised, "The whole point of going to the Fashion Design Institute was to obtain a great job. You skipped a step, and you no longer need the middleman. I will miss you personally and professionally, but if you were my daughter, I'd say go knock them dead." Cher waited in Bernie's office and became emotional while reflecting upon her rapid transition from a spoiled high school party girl, the reticent Gucci girl acting on behalf of a talking Pomeranian, to the serendipitous opportunity that awaited her arrival in Paris.

When Henri arrived, she wiped her tears and introduced the two clothiers. Blown away after meeting the number two man at the House of Dior, Bernie expressed to Henri what Cher had done for the store in design and marketing. He said, I'm sad to lose such a natural talent but proud to see her advancement. You will not regret your decision."

Cher felt uncomfortable packing for Paris the evening before her departure. Elise walked into the bedroom and placed a good face on an unpleasant situation. "Todd and I will miss you, but we understand your amazing chance to work with one of the top dress designers in the world. We wish you a safe journey and great success."

Chapter 21:

Gone Girl

George arrived with his driver at 7:00 a.m. to start their private eight-hour jet excursion from New York to Paris. Cher worried she had not packed enough clothes. George said, "You'll be the guest of a major fashion house. What you forget, they will make." His comment eased her tension.

Cher never thought she would be on a private jet destined for France. Perhaps counterintuitive, she felt safer on Citation X with George next to her than on a commercial flight. The airplane rolled into take-off position and made its wide turn onto the main runway. The Captain engaged the jet's full throttle, accompanied by the unnerving bumps and vibrations before liftoff. Cher loved every moment as the landscape below diminished in size.

Once the airplane reached its assigned altitude and cruising speed, the hostess handed them a breakfast menu. The food selections looked like those from a five-star restaurant.

After George ordered, he asked, "Now promise you don't have Mucky hidden in that massive purse." Cher looked puzzled by the question, and George reminded her of the last time their old buddy helped himself to George's meal on the train. "George, you know Mucky would have to be hog-tied to cross the Atlantic Ocean for eight hours even if God was the Chief Pilot."

They chose to have lunch at the bar, which was well stocked with Glenfiddich Scotch. Later, they both fell asleep during an airline pre-release of the movie *The Hangover*. When the two awoke, they had been in the air for four hours and had traveled to the Mid-Atlantic, and the hostess was preparing the aircraft for dinner. They had the choice of four cuisines. "George, I realize you like the train, but this is the way to travel. George corrected the Dior girl. "I didn't say that. William Gilmore did. I thought that was a good off-the-cuff answer at the time. You bought it." Cher said, "Very cute, funny man."

***

The pilot circled the airplane as it approached the Charles de Gaulle Airport. The cloudless sky expanded Cher's panoramic view of the celebrated lights of Paris and the famous star-bright Eiffel Tower, which withstood the war and became the focus of the historical city.

George arranged for a driver, who retrieved their luggage, and Cher's magical journey to Paris began. On this chilly autumn day, the Four Seasons remained George's hotel of choice. The art-deco hotel in the city's Golden Triangle, just off the historic Champs-Elysees, impressed Cher with its endless flower arrangements. "George, the aroma of the white, red, and purple flowers is amazing. The hotel looks more like a palace." George reserved adjoining suites, lavishly furnished, causing Cher to expand upon her lexicon of superlative adjectives. Cher said, "The assortment of flowers and their fragrance are amazing." There were more carnations, roses, and tulips of different colors in George and Cher's adjoining suites.

***

Henri arrived at the hotel to check on his guests. George welcomed him inside with a big smile. *"Bon après-midi mon ami.''* Henri smiled and said, *"Non, mon ami, c'est bonsoir. Tu as oublié le changement d'heure de six heures."* Cher felt like the odd person out. "Alright, you two, how about including me in the conversation? Nobody told me to bring an interpreter.'' George apologized and said, "We were discussing the time change. It might be wise for you to learn French."

They finished unpacking, and Henri said he needed fifteen minutes of their time and suggested a restaurant. "All the cafes are closed. Sémon's, a Mediterranean Meza restaurant in the Four Seasons, is open, and we can enjoy our privacy at this hour." Henri explained to Cher that Mezza is the Mediterranean equivalent of a Spanish Tapas restaurant. George and Henri ordered Moroccan Meatballs, Stuffed Portobello Mushrooms, Fried Feta Zucchini, and his usual Glenfiddich on the rocks. Henri had a Cognac, and Cher a Pinot Noir.

Henri's fifteen minutes turned into an hour, and Cher received a surprise job offer that overwhelmed her. She asked, 'Do you want me to prepare a second line? Do we have sufficient time?" Aware of the potential problem, Henri said, "You are more than capable. I've seen your portfolio and spoken to your school, Bernard Cushman, and those who have raved about your talent. Your fellow students say you are a remarkable designer with an intuitive marketing ability. After hearing your accolades, if we had not met, I would have imagined you had been in the business for decades."

Cher looked at George for advice after Henri wrote her proposed salary and bonus on the back of his card. Surprised by the news, George finished his drink in record time and remarked, "WOW!" With a sardonic smirk, Cher stated, "Thanks for that thorough opinion."

Cher suggested to Henri, "If the preliminary submission of our summer line is a bit delayed, we can issue an immediate press release explaining Dior's recent retention of a new and innovative fashion designer. We can add some intrigue by saying I do not wish to disclose my identity until after the show." Henri loved the idea. When they were ready to leave, Cher said, "George, be a dear and take care of the bill." When Henri wasn't looking, She stuck her tongue out at George and smiled at the surprised host for the evening.

At three in the morning, George heard the Dior girl stirring in the next suite. He knocked on the door and said with sympathy and humor, "You've been working all night. So you finished Henri's project." Cher chuckled, "Yah, right!" For the first time, George saw panic in her eyes. He grabbed her hand, and they sat on the couch. George said, "Class is now in session. After everything you've been through, I know you can do this. In Longboat, do you remember what I said? I consider you my peer. I meant it and wasn't blowing smoke. When I have a difficult task, I nap to reenergize myself. Then, whatever the project, I discount the opposition because I know I am better than them all. Establishing that mindset is crucial. I am my own competition and don't allow any second-guessing of my decision. If I don't do this, I box myself into a corner. Turn off the lights and go to bed. We'll talk more in the morning. You're starting on the inside gate, kiddo."

Chapter 22:

A Sketch is Worth More than a Thousand Words

George and Cher ordered room service for breakfast. She didn't want to eat, causing George to ask, "When you prepared your school sketches for New York Fashion Week, did you not sleep or eat?" George provided her answer and said, "Of course you did. So what is the difference now? A simple answer is you're playing with the pros. But one cannot pigeonhole an event. The objective of your present task is the same. You need to think of this job in simple terms. Your ability speaks for itself; you are the same person you were in New York. Henri is already impressed with your suggestion that the debut of the new designer's identity remains anonymous until the runway show ends. People love intrigue and speculation, and you improvised a keen way to give it to them." Cher's mood improved after the pep talk, and she attacked her breakfast.

She told George that while in school, she had visited the New York Museum of Fashion Technology for inspiration before making a single stroke on her sketchpad. "I'll do the same today. There is a famous Fashion Museum near us." George asked, "Do you mind if I join you?" Cher appreciated his support and company. "I'd love for you to join me and share your opinions about my ideas. What would I do without you?" George smiled and said, "Unlike last evening, you might have to pay for dinner."

Palais Galliera

***

The Dior girl continued her daily trips to the Paris Museum for the week. She reviewed her notes each evening and transformed them into preliminary sketches. Ideas began to flow. After following this week-long routine, she had many new designs to show Henri.

Chapter 23:
What About Mucky

The evening before Henri's arrival to see Cher's work, George and she ordered room service. The dress sketches were ready for George's opinion. Fanned on a dining table, Cher wanted George's verdict about her designs before dinner. He paged through the book without comment, sat down, and said, "I need to look at these again." After a more detailed examination, he said, "I may not know how Dior evaluates women's fashion. Still, for what it's worth if a woman dresses for a man's pleasure, my opinion is supported by the years of my experience admiring beautiful women wearing stunning gowns. Your sketchbook is amazing." Cher grabbed a bottle of Bordeaux, and the two toasted her work.

During dinner, Cher expressed concern about the consequences of a looming decision if Henri shared the same opinion about her work as George. "How do I manage Mucky if I formally accept a position at Dior?" George said, "Sometimes, the tough decisions we make can be hurtful to those we care about. I'll withhold my opinion of what you should do other than suggest the sooner the dialogue begins, the better clarity you'll have in making your decision. Email your sketches to Maria and Mucky, and let the discussion begin." Cher agreed with George, and after dinner, her dress designs took an electronic Trans-Atlantic journey to Longboat Key, Florida.

When George arrived at the hotel the following evening, Cher went to his room and said, "Henri loved my sketches. With some fine-tuning, he believes we'll be ready for Fashion Week in London and Paris."

When Maria received the sketches from Paris, Mucky meticulously examined the designs, loved the sketches, and exclaimed to Maria, "These are wonderful! He begged her to telephone Cher to congratulate her exquisite creation of impressive and groundbreaking designs. Maria explained, "Paris is six hours ahead of us. It's one o'clock on Wednesday morning. I don't want to wake them." Mucky wore Maria down with his unrelenting insistence, and they called.

When her telephone rang, Cher and George had just returned to the hotel from dinner with Henri. Before answering the telephone, Cher looked at George. with raised eyebrows, and asked, "Do you care to bet who's on the phone?" With the speaker engaged, Mucky learned the particulars of the Dior offer and said, "The good news is you got the job, but the bad news is I can't call you Gucci girl anymore." Cher explained she had little time to decide upon the offer sheet. Without realizing her acceptance meant a permanent relocation to Paris, Mucky encouraged her to accept the offer.

George told Maria to remove the phone from the speaker so he could warn her about a brewing problem. He said, "Maria, Mucky doesn't understand if Cher accepts the job, she'll have to relocate. I hate to put all of this on you, but would you explain this to him and say we'll call tomorrow? We don't want to debate the issue over a Trans-Atlantic call at one o'clock in the morning." Maria agreed, and George returned to the speaker to talk to Mucky. "Old buddy, it's sure great to hear from you guys. I told Maria I'd send you the offer by email in the morning. After you look at it, we'd like your opinion." George knew if Mucky believed they wanted his input, this would temporarily satisfy his wish to be in the vital decision-making loop. Cher and George wished both a good evening and said they looked forward to their next call.

***

Elise had a disappointing morning in Manhattan when Maria called to tell her about Cher's decision to accept Henri's offer and relocate to Paris. Elise asked about Mucky. "Is Mucky going with Cher? We would love to have him." The rejection forced Elise to swallow a second bitter pill. Maria said, "Cher felt Mucky would feel more comfortable with George and me, but there's no reason we can't keep in touch as often as possible. Once the weather warms, George and I always return to Manhattan. Cher would also be in Manhattan during New York's Fashion Week."

Elise broke the bad news to Todd when he returned from work. He said, "I knew this would happen. We're back to square one. Of course, Cher is my most important concern, but I will also miss that little dog."

***

An article grabbed Elise's attention while reading the *New York Times* feature page. Before bed, she told Todd, "Guess what I just found in the paper. There is an article about Pomeranians." After reading the story, she handed the paper to Todd and said, "It's a good piece about the breed, their intelligence, and their adaption to large cities." The story included a listing of well-known breeders in the area. The closest was a breeder in Sherburne, somewhere in Upstate New York. Elise copied the information. The breeder was Irving L. Muckelman, 104 Feitler, Sherburne, New York.

Chapter 24:
The Change of Heart

The excellent progress of Cher's preparation for the new Dior line permitted the fashion powerhouse to advance its schedule. Henri said, "My dear Cherie, I am confident we will be ready to register for Fashion Week in London and Paris." Paris followed London by two weeks, and the spacing between the events in mid-February could accommodate any mid-course tinkering before their Parisian runway appearance. Cher's success improved her budding confidence at work.

Unfortunately, her progress at Dior failed to transfer to her domestic problem with Mucky. As George had predicted, Mr. Muckelman backpedaled on his previous unequivocal approval for the new job after Maria explained the permanency of her move abroad. The Trans-Atlantic late evening telephone calls made no progress. Mucky rejected Cher's offer to join her in Paris. The result of their lengthy Trans-Atlantic phone battles only resulted in additional revenue for the telecom's overseas surcharges.

Mucky questioned why Cher could not work in New York. She explained, "I have to be in Paris. This is not a remote job. This bothers me as much as you, but I will see you during New York Fashion Week next September. Mucky rejected the proposal. "That's over a year from now! Why don't you just throw me a bone!" In the days to come, Cher knew the frustrating phone calls would not end, or as George

Markarian euphemistically stated, the daily dialogue would continue until Cher left for London.

***

When Cher registered at her London hotel, with Henri at her side, she told the desk clerk, "I'm embarrassed to ask this, but "I have been running for two days with little sleep. What's the day of the week?" He laughed and said, "Today, life moves quickly; I often hear that question. There is nothing to be embarrassed about. It's Monday, October 26."

***

The decision about Cher's permanent relocation, with Mucky remaining in Longboat Key, caused Elise to aimlessly mope about until she remembered the New York Times feature article about Pomeranian breeders. When Todd arrived home that evening, she asked, "Do you remember my mention of the *Times* article about Pomeranian breeders? I copied the name and address of an Upstate breeder. How would you feel if we got a little dog like Mucky?" Todd cherished his wife and asked, "If you want a dog and you think it would improve your mood, let's do it. There's nothing major on my schedule that James can't manage. We can go tomorrow. We'll enjoy the beautiful autumn drive." Cher kissed her husband and said, "I'll call and see if he is available tomorrow afternoon." Irving

Muckelman picked up the phone, agreed to meet them at three, and directed them to 104 Feitler.

Todd drove the Rolls. If the gas guzzler didn't have to stop for fuel, his heavy foot would have cut a half hour off the drive. The beautiful multi-colored leaves trumpeted the change of season. The contrast between the rolling hills of Up-State New York and Manhattan's mile-high skyscrapers seemed like they had entered a different world.

A roadside billboard welcomed the couple to Sherburne and advertised a luxury downtown hotel and a five-star steak house. Another billboard hyped a casino's upcoming debut. They stopped again for gas, browsed the convenience store, and double-checked their directions with the service station attendant. Main Street led to a bustling downtown business district. They checked into the ultra-modern hotel, ate lunch, and left for 104 Feitler.

New cell towers had been added, and they did not lose their telephone signal when leaving the highway. The new housing development offered modest half-acre home sites and had a model display home with a two-car garage temporarily converted into a sales office. Further down the paved two-lane street, survey crews marked the location of future homesites.

Chapter 25:

Sherburne-The All-American City

When they arrived at the end of Feitler Road, the homes were older and smaller. Todd opened his wife's door and noticed someone peeking from the front window. He told Elise, "Well, we know someone is home." Mr. Muckelman ran and welcomed the couple at the open doorway. He commented about their Rolls. "I once saw nothing but fancy cars like that at my brother's house in Longboat Key. I'm Irving L. Muckelman, but call me Mucky. Are you folks interested in a Pomeranian pup? They're great dogs, very smart cookies. Let's go to the barn and have a look."

Todd and Elise's amazing reaction was not lost on Mucky, who asked, "Is something wrong? You two looked like you were about to jump out of your shoes a moment ago." Elise said, "We have a close friend with a Pomeranian named Mucky. It's just an interesting coincidence that took us by surprise."

This coincidence also surprised Mr. Muckelman, causing him to ask, "Do you mind telling me who owns the Pomeranian?" Todd said, "Cherie Egegian, a former Fashion Design Institute student who lived with us in Manhattan until she accepted a position in Paris from The House of Dior. When she relocated, she gave Mucky to some old Longboat Key, Florida friends."

The discussion rekindled the memory of Mr. Muckelman's horrifying airline accident and the argument with George about whether his post-crash transformation to a Pomeranian was real or delusional. He had to be careful about what he said to Todd and his wife because he was aware of the ordeal experienced by George and Maria from their unintentional peek at the embedded secret of the Hereafter. He feared what would happen if he made a prohibited disclosure and speculated if it could result in his sudden death and his soul's annihilation from the October 29th airplane accident. Mr. Muckelman's decision to say nothing about the odd coincidence seemed the wisest choice.

They all enjoyed watching the tiny puppies push and shove their siblings for nourishment from their patient mother. Drifter, the only puppy given a name by Mucky, was the most aggressive of the pups. This caught the attention and amusement of the couple. Patiently waiting for Drifter to finish feeding, Todd picked him up. The puppy loved the attention when Elise cuddled him.

***

While the couple played with Drifter, Mr. Muckelman obsessed over whether his redeemed life would continue or another drastic change would occur tomorrow evening in a celestial replay of his flight to New York. He questioned if his salvation would last. Was a soul once saved, always saved, and could a Jew or anyone who was not a Christian ever be admitted to heaven?

***

Todd and Elise decided on Drifter. After they left with their new puppy, Mucky returned to the house. Sadie woke from her nap and sat at the kitchen table. She noticed her husband's concerned appearance and asked, "Didn't your people show?" Mucky said, "Yes, and they picked Drifter. I think they made a good choice. I saw a lot of love." Because of his glum look, Sadie wanted to know what bothered him. "Sadie, I knew the folks who bought Drifter. I met them during my transformation when Cher and I moved in with them. I'm confused about the correct sequence of time. Todd and Elise said Cher moved for a job at some fancy schmancy fashion house in Paris, and I'm living as a dog with George and Maria in Longboat Key. I am confused. I can't live in two places at once. I never told the young couple anything about this because of what happened to George and Maria when they came here looking for you. What does this mean for me? Either I am out of time here as a man or as a dog in Florida. Two of us can't exist at the same time. Who is the one out of sequence? Tomorrow is October 29th; what happens then?" Sadie kissed Mucky and said, "You will accomplish nothing with worry. You know whatever decision He makes will be right."

Mucky left for the living room. He decided to call Longboat Key and received a recorded message that the number was no longer in service. He received the same recording when he dialed the landline. He wondered if this meant Sherburne had the correct time sequence, which would mean the continuation of his contented life in Upstate New York.

Chapter 26:
The Defining Hour

Maria and Mucky greeted George when he arrived at the Sarasota International Airport from Paris. Unlike the reception he received at the Sarasota Train Station, Maria showed genuine affection toward him at the airport. It felt nice to be out of her doghouse. George left Paris at two a.m. early Wednesday morning, October 27th, and had gained six hours from his nine-hour flight when his jet approached the airport at five p.m.

Maria saw George holding a gift-wrapped box. He kissed her, gave her the gift, and said, "I bought this for you at Baunt. I hope you like it." Maria shopped at the famed Paris Jeweler before and said, "If it's from Baunt, I know I will." The gift was a stunning white gold, 9.50-carat diamond bracelet. After they arrived home, Mucky pestered George to tell him the price. The cost of $19,390 amazed him. Mucky complained, "I guess they had nothing you could buy for me?" George used the same line as he did on the train to Cher. "Isn't my presence sufficient?" Mucky gave a simple but blunt reply. "Of course not. Are you drunk? How many drinks did you chug on the plane?"

***

After unpacking and making some return telephone calls, they enjoyed a later-than-usual happy hour at the pool bar.

Maria's new bracelet had a mystifying sparkle during the setting sun. While enjoying their drinks, George spoke about Cher's new-found confidence at work. "It's unquestionable she has come into her own. Cher displayed confidence while expressing her ideas at Dior and accepted constructive suggestions without criticism. She trudged to a nearby library for a week and researched her new clothing line and learned how fashions correlated to world economies and events." Mucky agreed. "I always believed in her potential. She's one smart cookie."

They went inside at ten o'clock when the mosquitos started biting. Mucky laughed at them. "You guys should get this flea and tick collar I am wearing. Although not as nice as Maria's bracelet, it works against mosquitoes. I'm not bothered at all. Since we're going in, let's call Cher." George nixed Mucky's suggestion and reminded his old buddy about the time change. "I won't be the one to wake her at four in the morning."

***

THE SHERBURNE, NEW YORK MUCKY

Like a crow flying in a straight line north, twelve hundred miles from Longboat Key, the Sherburne Mucky worried if his life would be disrupted at 8:15 tomorrow evening. His non-stop pacing in the living room made a noticeable footpath, with his rhythmic repetition of, "Which one is out of time, Sherburne or Longboat Key?" Sadie became overwhelmed by his obsessive behavior, but unlike Mucky's airport drive for his brother's funeral, Sadie was not a captive listener. She chased her husband to the barn before her head exploded.

Mucky had spun out of control. When he again tried telephoning George Markarian, he received the same out-of-service recording. Mucky rejected calling Cher in Europe. Even if he successfully reached her, he reasoned Cher would only know Mr. Muckelman as a scruffy dog she last saw before leaving for Paris. A call from Sherburne, from a man she had never seen, claiming to be Irving L Muckelman, from a place she had never been, would be a disaster. He did not have the luxury of time to prove who he was, nor ironically have Cher as an accomplice to establish his identity as she did with George on the train.

Mr. Muckelman had to confront the effect of logical irrationality, the classical state of an inescapable paradox. The out-of-sequence timelines of separate events had developed independent scenarios of time and place.

He knew Jewish Scripture did not accept Jesus as the Messiah. If one did not believe in the Messiah, Jews could not rejoice in heaven because their sins could not be forgiven without their acceptance of Christ as their Savior. However, Mr. Muckelman took some solace in this because its interpretation lacked universal agreement among religious scholars. Because of this, Mucky prayed he qualified for the doctrine of 'once saved, always saved.' He felt saved from spiritual death and annihilation after the airplane accident on October 29, 2009. He was grateful for the opportunity to seek forgiveness and atonement from those he had wronged.

However, Mucky was afraid that if he was wrong, he would no longer be able to spend eternity with Sadie. He left the barn and returned to the house, apologizing to Sadie for upsetting her. He kissed her and promised to always love her. Mucky tightly held her hand and refused to let go."

Chapter 27:

What's Happened To Mucky And Cher?

Wednesday morning, Cher had breakfast in her room. Beginning at nine, she had a non-stop schedule with Henri at the fashion pavilion. A peek outside her window revealed a typical chilly, gloomy, foggy London morning. Cher hailed a taxi for 180 Strand Street. When she arrived at the venue, Henri and several assistants gathered around a conference table. The staff numbered and pinned each sketch to a giant corkboard. Pictures of the runway models were on the table and marked by random numbers. Dior assigned each runway model to the gowns they would debut during the show. The tedious process continued until two in the afternoon, and they did not finish planning for London's Fashion Week until five o'clock.

Confident they were well prepared for the presentation of their new line, Henri and Cher dodged the rain to have dinner at a nearby café. Henri asked if George had returned to the States. Cher explained her total focus had been on the show. "I haven't called or heard from him, but he should be home." After dinner, they ordered a bottle of wine and continued talking shop.

***

THE LONGBOAT KEY MUCKY—THURSDAY, OCTOBER 29TH

In Longboat Key, Maria went to bed early Thursday night. George and Mucky decided to watch television. At eight-thirty, George went to the kitchen for some chips, and when he returned, Mucky was gone. He woke Maria, and they looked throughout the house but could not find him. All the doors to the house were closed and locked. George said, "He just disappeared. I better call Cher." She did not answer the phone, and George left a brief message for her to return his call as soon as possible.

They continued to search for another hour with no success. George again tried to call Cher. He found it odd that the telephone recording had changed and now said the number was no longer in service. Maria walked into the room, looked at George's perplexed expression, and asked, "Didn't you reach Cher?" George threw up his hands in frustration. "Now we've lost them both. Cher's cell number is no longer in service. What the hell is going on?"

***

Henri Durand and Dior's Executive Designer, Elonne Cienne, waited for Cher's arrival at the Friday morning meeting. They waited until nine-thirty, called Cher's cell number, and received the same message as George. When Henri called her hotel, he received shocking news. The desk clerk said he found no record of Cher's registration at the London Rosewood Hotel.

Before dawn in Longboat Key, the Markarian telephone rang. Neither George nor Maria had gone to bed, and George immediately grabbed the phone on the first ring. Henri had called and explained, "Cher missed a nine o'clock meeting, and when I called her hotel, the desk clerk said she was not registered. George, I stood next to her when she checked in. I am at a loss of what to make of this." George told Henri to hold the line and told Maria, "Henri is on the line from London. Cher has disappeared as well." Returning to Henri, George said, "Maria and I are as perplexed as you. When we telephoned her, a recording said the number was no longer in service.

"We have two problems. Mucky is staying with us, and he has disappeared as well. Last night Maria went to bed early, and I watched some television. Mucky sat next to me on the couch. At about eight-thirty, I went to the kitchen for some snacks. When I returned five minutes later, Mucky was gone.

"He couldn't have left the house because all the doors were closed and locked. I woke Maria. We looked everywhere but could not find him. Doesn't it seem strange that they vanished at the same time?

Chapter 28:

The Crow's Return Flight to Sherburne

THE SHERBURNE, NEW YORK MUCKY

Sadie's hand turned numb. She pleaded, "Mucky, loosen your grip. I can't feel my fingers. It's midnight; let's go to bed." He walked with her to the bedroom with the stride of a condemned man. Mucky knew he would be unable to sleep. He changed into his night clothes and told his wife to go to bed while he sat on a bedroom chair, staring into space for answers. The only illumination came from a nightlight, with Sadie breaking the quiet of the night when she began to snore.

Mr. Muckelman glanced at his watch at two-thirty. The absence of answers to his concerns transitioned from fear to acknowledging a new introspection as he thought, *Lord, I've done the best I could. My attitude and acts of redemption came from my heart and were not done to please you. I pray you are not displeased with me because I did not and will never accept Mabel and other mean-spirited folks like her. However, I pledge to never use my hand to cause your creations to suffer. That is your job.*

A deep and echoing voice from the night startled Mucky but did not break the rhythm of Sadie's snoring. *Fear not, my son. Your Heavenly soul shall continue. The Heavenly power will sustain your faith and preserve what has been promised.*

Your transformation and the young woman who assisted were my conduits from heaven to determine if you had moved past your challenges for a successful redemption. They have served their purpose, and no sadness should result from their absence.

Mucky could not move from his chair. His Jewish religion would be no impediment to his eternal soul. As his creator, God cared about Mucky's heart and the heart of all his creations, not just external righteous deeds or the earthly scholars who often misinterpreted his Rule Book.

The moral of Mr. Muckelman's story was although he considered himself a devoted Jew and admired the Hasidic Jews of New York City, whom he described as 'those bearded people with long ponytails and black magic hats,' the Heavenly Father spared his soul from damnation, with the assistance of two celestial conduits. God had set a high hurdle for Mucky to satisfy the commitment to redemption. Both conduits had returned to their celestial world after allowing the Heavenly Father to determine a just and proper decision. There were no reincarnations, which had no existence in scripture, but they were God's celestial helpers.

Irving L. Muckelman's irrevocable redemption is contrary to Biblical scholars who have written a non-Christian Jew who does not believe in the Messiah can never attain such salvation. Some also consider salvation, once given, can be taken away. The Boss disagreed, and HE had the final say. Mucky's good heart won the day. There was also something about the boy HE liked but found difficult to express.

Chapter 29:

The New and Fascinating Drifter

During the summer of 2013, Todd and Elise Montgomery returned to Salt Lake City. With the Director's position filled, Todd accepted a demotion to Assistant Director. Because of their inherited wealth, the pay cut did not factor into the couple's decision.

Todd and Elise never saw Cher and Mucky after their mystical disappearance. Any mention of them only occurred during their occasional weekend visits to Mr. Muckelman's Sherburne farm so Drifter could run wild in the pasture.

Soon after their unsolved disappearance, Todd and Elise flew to Sherburne with Drifter. While playing fetch with the energetic dog, Todd told Mr. Muckelman about his investigation and said they could not find a single lead in either case. "With all my years in law enforcement, this case is the strangest I have ever encountered." Mucky was bursting at the seams to tell Todd why the cases would never be solved, but he feared the consequences of any disclosure.

***

During a crisp autumn day in 2013, Elise and Todd received terrible news from Sherburne. Irving L. Muckelman had died after falling off a tractor while clearing the foliage for next year's crop. Sadie called Todd and Elise with the sad news. They consoled Sadie and planned to attend the funeral. However, Sadie had called them to explicitly advise them that no funeral had been planned nor needed.

Sadie knew The Heavenly Father had honored the permanent salvation of Mucky's eternal spirit. Thereby, the earthly ritual of a funeral could be eliminated. She did not want Elise and Todd to encounter the 2019 emotional turmoil endured by George and Maria when they came to Sherburne searching for her and the shocking revelation in 2029 when George returned alone. After the telephone call, Todd noted Sadie's unusual tone and demeanor.

***

A month following Mucky's fatal accident, Todd scheduled a meeting at the Manhattan FBI Bureau. Elise and Drifter joined him on the flight for a planned visit with Sadie in Sherburne.

When they arrived at her house, Sadie embraced them and said, "How wonderful to see you all. Let's march inside and visit."

While reminiscing about Mucky, Drifter became restless during their chat, and Todd took him outside to play. Although Drifter seldom barked, the dog became unsettled for no apparent reason when he and Todd walked to play in the pasture. Mucky's spirit had arrived from the barn to watch the two play, and Drifter saw him and continued to bark and howl until Mucky spoke to the spooked dog. *Drifter, it's Mucky. I raised you. It's all okay. Everything is fine; it's time to play fetch.* The dog calmed down. Todd didn't understand what upset the dog, nor what calmed him.

***

Sadie spoke to Elise about Drifter and said, "I think Mucky favored him. He laughed when he told me how Drifter would muscle his way to his mother to be the first to feed." Elise asked, "Was he otherwise different from the rest of the litter?" Sadie explained Mucky oversaw the dog business. "I rarely entered the barn. But Mucky did discuss some of his distinctive traits. I know you picked the best pup. Mucky always believed he was the most attentive and curious of the litter. Before Mucky fenced in the pups, Drifter had wandered away from the mother, found a black Angus cow, and studied the animal as if preparing to draw its portrait. Other animals have a short attention span, but Drifter always kept his focus."

Elise mentioned Drifter's new sleeping habits. "Sadie, when Drifter is in a deep sleep, his loud snoring causes Todd to wear earplugs to bed. A few months ago, Drifter woke me, and I caught him in a dream. I saw his legs moving as if being chased. Drifter's dreamlike growling sounded like cursing. I woke Todd to tell him what I saw, and he made two flippant comments. 'If there's more cursing, wash out his mouth with soap, and please don't wake me unless Drifter starts reciting The Pledge of Allegiance.' He thought I was nuts." Sadie laughed at the story and said, "Well, I guess you're stuck with a cursing dog." The two smiled as Todd returned to the house with Drifter.

***

Before bed, Mucky told Sadie that Drifter saw his spirit and started howling while playing in the pasture. "He remembered me, and I had to calm him before he paid attention to Todd. He sure is a good dog." Sadie amplified her concern about Drifter's behavior. "Mucky, that bothers me because while Elise and I were chatting on the porch, she told me she thought she heard Drifter talk. She said it sounded like the dog was cursing during a dream. What happens if Elise is right and the dog begins to talk and discloses what he knows? I often needlessly worry too much, but I thought you should know." Mucky said, "I know the Boss will figure it out. After all, you can see and converse with my spirit. One should not question His judgment."

Chapter 30:
Welcome to the Family

Elise's identical twin sister, Vanessa Martin, lived near Big Sky, Montana. She automatically became eligible for a ski club membership when she built her luxury home on a Yellow Stone Ski Club lot. The exclusive club required yearly dues of $400,000. Todd and Elise encouraged her to join and agreed to finance her membership.

She had arrived in Salt Lake City the day before Todd and Elise's return from Sherburne with Drifter and planned to meet them at the airport. When she attempted to enter the private aircraft gate, an airport official stopped her. However, the young security official succumbed to Vanessa's manipulative and flirtatious manner and allowed her passage with a smile.

Vanessa was also a fundamentalist Mormon and had agreed to a plural marriage with Todd. They were always fond of each other, and Elise supported and encouraged the Morman Sealing of their nuptials. A plural marriage to Todd's sister-in-law had the benefit of preserving the secrecy of their union because the living arrangements of Elise's sister could easily be explained. The civil ceremony, planned during the New Year Holiday, included a secluded skiing honeymoon at the Yellow Stone Ski Club.

When Todd and Elise arrived from Sherburne, the identical sisters wanted to see how Drifter reacted to them. They sat facing each other on separate couches and simultaneously called for the dog. Drifter hesitated, refused both commands, and walked to the bedroom for a nap. Todd said, "You are definitely sisters; you confused the poor dog." Although Todd had been skeptical of his wife's claim, he said, "When Drifter wakes, be prepared for some amazing bad language." Vanessa laughed at the remark and said, "Todd, honey, You make no sense." Todd facetiously replied, "Do you mean to tell me your dear sister hasn't told you about our sleeping dog who swears in his sleep? Vanessa said, "Only after I hear such nonsense will I believe it."

Thirty minutes later, they could hear Drifter snoring. Vanessa said, "Well, the dog didn't curse, but he certainly is a prolific snorer," Vanessa laughed at the sleeping animal and said, "What incredible noise for such a small dog."

Excited about their New Year's marriage plans, Todd, Elise, and Vanessa decided to include Drifter during their ski holiday. They planned to celebrate and cherish the holiday, Todd's marriage to Venessa, and welcome Drifter to their new family.

Venessa asked Todd if she should sell her Montana house. "It is a 400-mile trip from here. Will we use it often enough to justify the expense?" Todd did not equivocate about his wish to keep the house and club Membership. "Uncle Bill's inheritance makes us economically independent, and my job doesn't pay peanuts. Definitely keep it."

Their loud banter and laughter in the living room woke the napping animal. From the bedroom, they all heard the dog's complaint about the noise, "When I bark, they carp, but they make all that racket when I'm sleeping." Drifter slowly walked into the living room, looked at Elise and Vanessa, jumped on Vanessa's couch, and said, "You're Vanessa. You can't fool me." Vanessa blacked out and recovered after Drifter licked her face. Drifter continued, "I'm not surprised I can talk because when I was a puppy, Mr. Muckelman said he used to be a dog and that someday I might talk. I miss him. Can I go outside?"

When everyone's jaw recovered, Todd said, "This is all absurd. Mr. Muckelman was once a dog?" Elise, silent throughout the episode, quipped, "Why would the breeder lie to a dog?" Still not fully recovered, Vanessa mumbled, "What the hell is going on!"

***

Todd accommodated his future wife with Sherburne's back story. "When Elise and I were introduced to Drifter's breeder, he told us to call him Mucky. Noticing our unusual reaction, he asked if we were okay.

"I told him we had a friend who owned a Pomeranian with the same name, and the dog and its owner lived with us before she moved to Paris as a designer with the House of Dior. The dog stayed in Florida with her friends George Markarian and his girlfriend Maria. As the Field Director of the Manhattan FBI office, we investigated the disappearance of the dog's owner in Europe and her dog in Longboat Key, Florida. Both disappeared on the same day. We never discovered what happened to them, and the case remains open. We have three additional parties to interview, George

Markarian, his girlfriend, Maria Allison, and the deceased breeder's wife, Sadie Muckelman.

Chapter 31:

Fewer Answers to Growing Questions

Drifter's newfound ability to speak caught everyone off guard and disrupted Todd's sleep even more. His earplugs proved useless on the evening he revealed his secret to Vanessa.

Rather than lay awake in bed, Todd went to his office to plan his interview with George and Maria. After some time, he decided to use a narrative approach instead of a Q&A format.

Todd managed a few hours of sleep and called Longboat Key, Florida. George answered the phone. Todd said, "Something new has developed in the missing cases of Cheri Egegian and her dog. I can fly out tomorrow if it's convenient." George told him to call with his arrival time, and he would meet him at the airport.

***

During George's drive from the Sarasota Airport, Todd said, "Elise and I miss Mucky and Cher. The two can never be replaced. We bought a Pomeranian puppy like Mucky from an older man in Sherburne, New York. The *New York Times* rated him as one of the best Pomeranian breeders in our area." George didn't hear a word of Todd's conversation. What he saw in his rearview mirror concerned him, and he said, "Damn, we're being pulled over." Todd said, "Let me talk to the officer. How fast were you going?" George grinned and said, "Only one hundred and ten."

George pulled over, and the officer routinely called in the plates and his location. The answer to his radio call took a while, and George asked Todd, "G-man, you said you would take care of this. Why are you sitting here?" Todd explained, "If I jump out of the car while the officer runs a check, he'll get spooked. I don't want a gun pointed at me. Sit with both hands on the wheel until he comes to your car."

When the officer approached George, Todd showed him his FBI identification. They walked behind the car, and Todd formally introduced himself while the policeman took a closer look at his FBI credentials. Todd explained they were investigating a case from the Director's Office of Investigations, and a new lead surfaced in a dormant abduction case. Todd said, "The driver is a key witness and an acquaintance. We were rushing to Longboat Key to speak to his wife." The officer returned George's license and said, "Tell the driver to slow down." When Todd handed George his license, he said, "Good work! I'm going to keep you around." George lowered his speed to one hundred.

***

Maria heard the roar of George's car and ran outside to greet Todd with a hug and said, "It's great to see you again; please give Elise our best." George told Maria, "Todd saved me from a one hundred and ten speeding ticket. Do you think I can keep him?" Maria quipped, George, you know you would never take care of him."

George asked, "What's the new information? Todd sighed and said, "What I'm about to tell you is very strange, and I assure you, I have not lost my mind. Elise read a feature story in the Sunday New York Times about the top Pomeranian breeders in our area. We chose a breeder from Sherburne, New York, named Irving Muckelman. He asked us to call him Mucky. From our shocked appearance, Muckelman thought something was wrong. We explained the coincidence of a friend's Pomeranian with the same name. He asked for the owner's name. I usually refuse to disclose names to strangers, but I gave him Cher's name. He laughed and said, "Now I understand your reaction. I knew them and read they were missing."

 Mr. Muckelman did not disclose the whole story. But he came close to crossing the line when he looked intently at the couple and asked, "You two look mighty familiar. Have we ever met?"

Todd asked George and Maria, "Did you know a man named Irving Muckelman?" George and Maria looked at each other with amazement. George said, "Yes, we did, but before we go any further, we assure you we have not lost our minds either.

"I'm pouring us some good scotch straight up. With what we're about to discuss, I don't want to hear about the Mormon temperance rules.

"Maria and I initially thought Irving Muckelman perished during an October 29th 8:15 evening flight home to New York City. He had been a guest in our house for the funeral of his billionaire brother, my business partner Saul, who left him ten million dollars. Before he left for home, he established a charitable educational foundation, and I became the Executive Director.

"We were devastated by Irving Muckelman's death and had been slowly recovering from the tragedy until we received a St. Louis telephone call from Cherie Egegian, who claimed she knew him. She called about receiving a Muckelman Foundation grant to the New York Fashion Design Institute. Cher's story made no sense, for she claimed to have spoken to Mr. Muckelman after his death.

"I arranged train transportation from St. Louis to my Longboat Key home. Cher insisted that the dog accompany her. I secretly boarded the train in St. Louis to expose her as a fraud.

"During the trip, Cher related personal information that only Mr. Muckelman, Maria, and I knew. She claimed her source of the knowledge came from Mr. Muckelman, who had been transformed into a communicative dog she adopted. The transformation to a Pomeranian, which only she would later hear, and relocation to St. Louis occurred after Mr. Muckelman's fatal airplane accident on his return flight to New York.

"After Maria and I acceded to her mystical claim, we also could freely converse with the animal." Todd asked, "Do you believe there is any connection between their disappearance and the anniversary date of the airline accident?" "Perhaps," Maria said. "There may be a religious connection. I researched the internet and found an incredible amount of religious material that convinced me their disappearance was a spiritual clean-up action. By that, I mean Cher and her dog may have been religious conduits that saved Mr. Muckelman from certain death and the annihilation of his soul after the airline disaster.

"The NTSB said the flight manifest listed 189 aboard, but they only recovered 188 bodies, presumably Mr. Muckelman's missing body. The agency could not explain the discrepancy. His transformation into a dog allowed him a limited and difficult time to seek redemption. He returned as an earthly man once the Redeemer determined he warranted salvation. There could not be two Muckelmans, nor would there be any further need for Cher to function as his medium for communication. After the transformed dog and girl served their purpose, they were returned as spirits. Todd, this is the only logical explanation George and I can accept. I doubt anyone can construct any better explanation for these mystical events." George added the dog breeder knows this because while a dog, Mucky told me he worried God would rescind his salvation and his body would be found at the crash site."

Todd grew weary of the convoluted details and said, "We've been at this all day. If I get a hotel, can we continue this tomorrow?" George gladly agreed. "Of course. We are as curious about this as you but forget about the hotel. You're staying with us. Use our telephone to call Elise."

Chapter 32:

Just the Facts

All the crazy and jumbled facts dancing in Todd's overworked mind did not prevent him from sleeping through the night. After breakfast, everyone returned to the living room. They would continue with their second session in their night clothes and robes.

Todd thanked his hosts for putting him up for the night and said, "I woke up troubled about Mr. Muckelman's candor about Cher and her dog. Maria, your theory of a religious conduit is a reasonable explanation for their disappearance. After finishing here, I plan to visit Sherburne to speak with Sadie.

"I'd like your input on Drifter. As discussed yesterday, we bought the dog from Mr. Muckelman, and we learned three days ago Drifter could talk. Based on what the dog said, I'd like your opinion on whether this is connected to the disappearance or abduction of Mucky and Cher. Also, Drifter's ability to talk coincided with the disappearance of Cher and her dog, Mucky."

Todd related what Drifter said. "Before Drifter could talk, Mucky would check on the litter and find Drifter had wandered away from his mother. When Mr. Muckelman found him, he told the puppy about his transformation from a man to a dog and his ability to talk and understand people. Mr. Muckelman told Drifter someday he might also be able to talk. I learned much from Drifter that could help us understand what occurred." George and Maria had no thoughts of whether Todd's dog had any relationship with the investigation.

***

George called Sadie Muckelman while driving to the Sarasota Airport. Like he usually did while alive, Mucky's spirit spooked the kitchen when Todd called, and he asked, "Who called?" Sadie said Todd Montgomery, and she wasn't pleased. "I told you that FBI agent would cause us trouble. He didn't tell me what he wanted, but I bet it's about that dog you sold him. I knew there was something about him I didn't like. He'll be here Wednesday morning at ten." Mucky's lack of concern soon changed.

***

During the middle of the night, a celestial messenger appeared before Mr. Muckelman in the usual manner. Mabel continued to sleep as Mucky listened to the spirit who said Drifter was more problematic than Mucky thought.

You spoke to this animal and disclosed embedded secrets during your visits to the barn. Drifter is talking and has told the FBI agent the history of your plight for redemption and the reason you survived the doomed flight home. He plans to visit your wife. Mr. Montgomery suspects you were not forthcoming when he spoke to you about the missing woman who adopted you in St. Louis. He wants Sadie to talk and sign a written statement confirming what you told Drifter. He can't credibly base his investigation only upon a talking dog, as the former United States Attorney General will learn. There must be no confirmation!

You will suffer no consequences for your breach to the pup because you were unaware he could understand what you said and could later disclose what he heard.

Sadie is his only credible witness. You must deal with Mr. Montgomery to avoid serious consequences that could follow. Advise Sadie not to speak and reveal something she would regret.

***

During breakfast, Mucky told Sadie, "You tell Todd you are not emotionally able to talk about my death or what anyone may have said about my transformation and disclosures to Drifter. Todd will come to a dead end in his investigation, and we will be done with him."

***

When Todd arrived in Sherburne, he noticed Sadie's frosty reception, sharply contrasting his previous December 11th visit. She poured two cups of coffee, and Mucky joined the two in the kitchen. He reminded Sadie, *Don't disclose anything. Explain you are too distraught with my death to answer his questions.*

Todd asked how Sadie was coping, and she gave a perfect answer. "All spouses with good marriages have a difficult time, but when your husband dies from a preventable and freak accident, the loss is more difficult to overcome." Todd correctly presumed he had made a wasted trip. But he continued, "Sadie, what do you know about Mr. Muckelman's survival from the October 29, 2009, airplane accident?"

Sadie responded as directed, "Todd, you're a lovely man who must do your job, but I can't discuss this anymore. It's too upsetting. I'm sorry you came all this way. Why don't you stay for lunch?" Todd declined and said he had too much work. He kissed her and told her to call if she needed anything or changed her mind about talking.

Chapter 33:

Frustration

Because of his lack of success with Sadie, Todd's flight home seemed longer than usual. He considered convening a Grand Jury that would compel the answers to his questions, but he could not be so callous to a widow. Additionally, he considered the odd nature of the case and worried he might be laughed out of town. Todd decided to terminate the investigation when he returned to Salt Lake City.

When Todd arrived home a day early, Elise and Vanessa greeted him and noticed from his behavior that his trip had not gone well. The women told him Drifter still talked, but what he said had no significance to the investigation. Todd said, "It's time to close this case and move on. At least people won't think we've all lost our minds. Everything we heard from Drifter is never to be repeated." The women agreed and were glad the case would be closed. Vanessa said, "Sometimes the events one questions are not meant to be understood."

Todd called the office and said he had just arrived home and would be at work early tomorrow. Sadie and Mucky were home free. Todd, Elise, and Vanessa had a late dinner and retired early. The women were right; Drifter had nothing to say except to welcome Todd home with many licks of love.

***

In their bedroom, Todd asked Elise, "When did your sister become a philosopher, saying some questions aren't meant to be answered?" Elise laughed, saying, "You should have asked her to cite some authority."

Vanessa Martin, although flirtatious and considered a beautiful airhead, graduated cum laude from Yale University. After four years of living off-campus in New Haven, Connecticut, she became what Elise called a free thinker, someone whose ideas and opinions often strayed from mainstream thought. The two sisters were physically identical in every aspect. However, their personalities differed. This pleased Todd because this was the only way he could distinguish between the two women.

Vanessa had many boyfriends but had never married. When Todd collected sufficient nerve to ask her to marry, she accepted. She always wanted children and a big family and feared she might reach the age where her hope for offspring would fade with age.

***

The following morning, Todd left for the FBI headquarters at seven. It felt nice to be in familiar surroundings. A sealed envelope marked VIP contained a memo from Salt Lake City's FBI Director, Henry Schwartz. It read, "Please immediately read a classified file in the safe. The office received the report yesterday from the Homeland Security Director." While Todd read the documents, he kept repeating, "Holy Shit! This guy wants to blow up and radiate the Hoover Dam!"

Chapter 34:

No Rest for the Good Guys

When the Director, Henry Schwartz, arrived at Todd's office, he saw him writing notes from the classified file. Todd said, "We have Rex Harper in Federal Detention and have confirmed he is a forty-two-year-old single male, never married, and still living in the home of his deceased mother's run-down neighborhood. He is an unemployed construction laborer. Security stopped Harper when they caught him taking measurements of the thickness of the dam's walls and making notations in a folder. As they approached him, he made a run for his car. The police arrested and cuffed him. What the hell was he doing, or how did he explain his possession of measurement equipment, cameras, micrometers, and survey equipment?"

Henry explained, "He didn't. He clammed up and requested an attorney. He has a ten-year-old felony conviction for bank robbery and served time in the Menard Federal Penitentiary. His parole recently ended. He drives a new Lincoln. You're looking at the diary the police found in plain view on the passenger seat when we impounded the car."

With a grim look, Todd said, "If he can accomplish what's in this diary, we could have something worse than nine-eleven. He can't do anything while in detention, but the diary doesn't foreclose the existence of any associates who could. We have to move fast on this one. Can you arrange a meeting with the people at the Hoover Dam and find a top-notch engineer?"

Henry Schwartz had been on the same wavelength as Todd and said, "I have an engineer set to meet with us tonight. Tomorrow, we'll fly to the Hoover Dam and speak to the chief security officer and the officials in charge." Todd called home and said he had an important meeting at the office and would be arriving home late.

Henry arranged for three engineering specialists to attend the evening meeting: Hydraulic, electrical, and mechanical.

***

During the evening meeting with the engineering specialists, Todd wasted no time summarizing the grim description of what was planned by the detainee. "I hope everyone has eaten dinner because, if not, you will have no appetite when I finish.

"We seized a diary in plain view after an inventory search of a suspect's car that detailed a plan to blow the Hoover Dam and radiate the water," he shockingly told his audience. Some writers have contended that the thickness of the target's steel-reinforced concrete walls could withstand a direct hit from a 747 jet. Perhaps this is why the plan to radiate the water may be a second-stage alternative." Todd asked the engineers, "Let's first assume the dam walls are fallible. What would result if the walls were blown?"

After conferring, the engineers all agreed the area would suffer fatalities from extreme torrents of water similar to a tsunami and an absence of drinkable water. The hydraulic engineer took the extreme position that people in the area could theoretically have a short walk to the Pacific Ocean.

The engineers predicted mass panic, a looming world economic disaster, and a probable depression in the United States. The radiated water would alone be catastrophic even if the dam walls held.

***

The following morning, Henry and Todd arrived at the Hoover Dam. Top security officials and executives of the iconic structure greeted them. In a closed conference room, Henry and Todd explained the threat. After the briefing, the chief security officer's face lost its color. The Hoover Dam officials said they reviewed a security analysis and summary report completed six months ago. Henry said, "We will review your report in greater detail with the engineers we have retained, but we think prudence requires a new review as soon as possible to account for the current threat. For example, I may have missed this in the recent security report, but there is no mention of tainting the water with radiation and the resulting consequences."

Henry and Todd's secured beepers simultaneously sounded. The Salt Lake City office informed them that an emergency bond hearing had been set for Rex Harper on Monday, December 23rd. The conference ended with a unanimous decision to suspend public dam tours indefinitely. The Federal Aviation Administration established a no-fly zone within a two-hundred-mile radius of the dam.

***

When the parties arrived at the Federal Courthouse Monday, the Court Clerk said, "Judge Proud would like to see all counsel in chambers before we begin." Proud set the emergency hearing after reading the government memorandum requesting no bail. The United States Attorney argued his reasons before the Judge in chambers with a court reporter present. He presented the diary's contents and produced the engineering affidavits of the opinions of the consequences if such a horrific plan succeeded.

After 43 years on the federal bench, Judge Proud usually used his law clerks for legal research and writing, but the Harvard-educated Judge researched and analyzed this case himself. After the attorneys had finished their remarks, the judge went off the record and allowed the court reporter to leave.

Judge Proud explained, "I wanted to go off the record. I am ordering a temporary gag order. I want to be frank with the government." As Henry and Todd listened, the Judge told the United States Attorney, "You may have some problems with your case. The Court has read the defendant's diary, and there is no question it details a dreadful act of terrorism. However, no overt acts have been alleged. The diary fails to mention any accomplices or actions that would constitute an illegal conspiracy to commit a crime. What we have is a recitation in a diary. Suppose we prosecuted everyone who considered committing a crime without actually carrying out any overt acts in furtherance of the unlawful operation.

Wouldn't we violate the Constitution and run afoul of the First Amendment's guarantee of free speech?–Okay, with that being said, let's return to open Court."

The Judge read the charge to the defendant and asked for his plea. The Federal Public Defender entered a plea of not guilty, waived the Information's reading, and asked that his client be placed on a recognizance bond, promising to appear at all future Court settings without posting a cash bond. The United States Attorney requested that no bond be set or, in the alternative, the Court set a bond of two million dollars and a surrender of the defendant's passport. The Judge ruled, "Based upon our discussion in chambers, the Court sets the defendant, Rex Harper's bond at $10,000, surrender of his passport, and release from custody with posting 10% cash, in lieu of the ordered bond. If there is nothing further, the Court stands in recess." The defendant shocked his attorney when he blurted to the Judge, "I have the $1000, but because of this, my 11-year-old nephew, who stays with me, lost his dog. What will the government do about that?" Judge Proud told him to confer with his counsel, banged the gavel, and returned to his chambers.

Todd Immediately ordered around-the-clock surveillance of the defendant's residence. If anyone entered or left the house, the agents logged the activity, followed anyone leaving, and changed agency cars every three blocks.

Rex Harper's nephew and the lost dog gave Todd a unique idea. If Drifter could access the house, his presence would be as valuable as having an agent present to hear every conversation in the Harper home. During Todd's return to the office, various scenarios raced through his mind. He first needed confirmation of the nephew's lost dog.

A home across the street from the suspect's house did not have a for-sale sign in the front yard for long. Two FBI agents posing as a young married couple, purchased the home, which became the agency's new surveillance facility.

With the surveillance in place, a young and attractive female agent noticed a child from the targeted house playing alone in the front yard. She walked across the street, introduced herself to the child, and said, "Hello, I'm Nancy Adams." The polite twelve-year-old boy introduced himself, "I'm Jessie Harper." Agent Adams continued, "My husband and I moved into our new house across the street. "I'd like to meet your mom and dad." The boy said, "I live alone with my uncle. He's still downtown." Jessie Harper appeared to be a polite and well-balanced child and did not appear to be tainted by the sociopathic behavior of his uncle.

 The two continued to chat, and the subject turned to pets. Jessie said, "I used to have a dog, but he ran away." The agent responded with sympathy, "I am so sorry about your dog. However, we may be able to solve two problems at once. My brother just got married, and his wife is extremely allergic to animal fur. We must find a new home for the little Pomeranian. If your uncle agrees, I can bring him over, and if you like Drifter, the dog is yours." The elated youngster thanked her and said, "You're the best! What a great neighbor!"

Agent Adams returned to the FBI surveillance HQ, telephoned Todd, and said, "Drifter is in. The uncle just called his nephew and gave his approval." Agent Adams asked, "How will the dog relate any information to us?" Todd hesitated and said he had developed a special wire for the dog.

Chapter 35:

As One Bank's Vault Closes, Another Opens

Rex Harper posted bond and returned home at six o'clock. Before making dinner, he telephoned his crew in the room where Drifter and his nephew were playing. The uncle suspected his telephone line had been tapped and had privately discussed with his crew of conspirators how to use the tap to their advantage. Drifter closely listened to Harper's end of the call as he said, "I have a broken water pipe in the basement. The place is flooded. I may have to postpone our bank operation. I will call you later this evening with an update."

Harper popped two frozen meals into the microwave. While eating their gourmet dinner, Harper told Jessie he had business in Houston, and his Aunt Dorris agreed to take him for a few days.

Later in the evening, Harper used his tapped phone as his proxy to transmit misinformation to the FBI and called his crew as promised. "Guys, I have made no progress in the basement. I'm afraid we will have to postpone our hit on the downtown Bank of America. We'll celebrate Christmas night with plenty of money." Harper could not have accomplished his objective much better without requesting an RSVP from the FBI.

When Drifter scratched at the front door, Harper told his nephew, "Let the damn dog out before he pisses on the floor." The dog ran across the street, and the FBI agent, who drew the short nightshift straw, saw Drifter and called Todd, who directed him to keep the dog inside. When Todd reached the FBI's surveillance house, he and Drifter left for another room. Todd asked his unique sleuth, "How much time do we have?" Drifter said, "I am sure Hoover Dam is not a target. Listen to the tape."

Todd had to play the tape several times because he lacked the luxury of having the distortions removed by FBI analysts. Drifter said, "I clearly heard him on the telephone, and there is no doubt they plan to hit the downtown Bank of America on Christmas. Harper arranged for Jessie and me to spend a few nights with his aunt, whose house is a block away." Todd laughed and told Drifter, "Well, go back and have fun at your sleepover."

***

Judge Proud's prescient comment proved to be correct. The confiscated diary did not foretell the Hoover Dam's cataclysmic blowing and its water's radiation. A less devastating crime had been planned.

The Harper gang had learned from a Bank of America co-conspirator about a Tuesday, December 24th Federal Reserve delivery of a large amount of currency to the targeted downtown Salt Lake City bank.

All large banks needed additional liquidity for their branches during Christmas because of customer cash demands. On Tuesday morning, an armored car received a Bank of America counter-signed receipt for Federal Reserve funds of twenty-three million dollars.

***

Monday evening, Todd called Director Schwartz and informed him of the Bank of America heist. The Director arrived at the office with his secretary at 9:00 p.m. After Schwartz listened to the intercepted calls and learned about the confidential informant, he told Todd to dictate the government's probable cause statement to his secretary. Henry said, "I'm certain Judge Proud will issue arrest warrants based upon the exigencies of time and your Affidavit attesting to the reliability of your confidential informant." Henry left to telephone Judge Proud's home.

Todd realized a severe problem could arise if the identity of his informant ever became known. Before dictating the sworn statement, he understood his potential prosecution for perjury, the loss of his law enforcement career, and the perpetrators walking free, but he had no other options.

***

Director Schwartz reached Judge Proud and drove to his home. When he told the Judge the suspects had targeted the

Downtown Bank of America and not Hoover Dam, Proud sighed with relief. However, Henry said, "I'm still bothered about the iconic dam's vulnerability in the future. Judge Proud agreed and said, "Next time you're in Washington, see if Homeland Security is on the ball about such a potential attack. I'd appreciate your opinion."

After reviewing the prepared documents, the Judge issued the warrants. Schwartz called Todd from his car and said, "We've got the green like. It's your operation. Round up the bastards." For a deterrent before executing the warrants, the Director arranged for five Salt Lake City police units to park at the bank all day and night during Christmas, with their emergency light bars flashing.

On December 25th, five tactical units consisting of thirty paramilitary-trained FBI Agents assembled at headquarters within twelve minutes of receiving Todd's call. Todd established a crime scene perimeter for the house. The Salt Lake City Police were responsible for barricading the streets and logging everyone permitted into the police zone.

***

Todd peered through the window of the agency's makeshift surveillance house to watch his well-armed paramilitary team swarm the Harper home. After they rammed the door off its hinges, they found the house empty. Harper's new Lincoln remained in the driveway. The unsuccessful raid angered and baffled Todd. He couldn't understand how the suspects eluded the FBI's surveillance team, with additional agents posing as linemen perched on a telephone pole high above the Harper home.

***

Todd spent a restless night at the agency's house. He awoke at 4:00 a.m. Christmas morning. He glanced at his watch at 4:30, and seconds later, a powerful explosion caused a massive fireball and shook the FBI surveillance house. Todd yelled, "What the hell was that?"

 On December 25 at 4:30 a.m., a powerful bomb had detonated at the Chase Bank, one block from Bank of America. The Bureau had no intelligence about Chase's inclusion in the operation.

Although stationed at Bank of America, the explosion prevented the police from responding. The officers had to dodge missiles of sharded glass as the concussion blew out the bank's windows. Additional units of FBI, Salt Lake City police, firefighters, and paramedics raced to the explosion site.

The FBI bomb squad, wearing protective gear, was the first to enter the targeted bank. After they cleared the building for explosives, the Bank's President opened the vault and saw the Federal Reserve currency bags on the transfer cart from the Federal Reserve delivery. After checking the bags, the agents found piles of fake money. The bank's alarm had not been compromised, and there was no evidence of forced entry. The twenty-three million dollars of currency had disappeared. Houdini could not have performed the feat any better.

Chapter 36:
The Great Magic Act

Although the Montgomery home was festively decorated for Christmas, the holiday spirit fizzled when Todd returned home at 10:00 a.m. Christmas day. He had been up for more than twenty-four hours and had not eaten or slept. The women saw his anger and frustration. While they sat in the living room, he displayed uncharacteristic black humor.

"I have good news and bad news. First, the good news is the Hoover Dam will see another day. The bad news is we were outsmarted and embarrassed. The crooks performed a magic act Las Vegas would love. They managed to steal and escape with twenty-three million dollars of the Federal Reserve money delivered to the downtown Bank of America Tuesday morning, and somehow, they managed to do so without entering the bank. We knew their names and had them under constant surveillance. They managed to disappear with the money and added insult to our embarrassment by blowing up the downtown Chase bank. Perhaps it was a diversion, or they wanted to celebrate their victory with toasted marshmallows from the raging fire at the bank." Todd rose from his chair, walked to his office, and shut his door.

Elise and Vanessa felt awful. They wanted to help. Elise said, "Probably we should just leave him alone. He hates to be patronized with sympathy. He will think this out and find the answer. I wonder when Drifter will be returned?"

***

While jotting some notes, Todd sat at his desk, with the theft eating away at his psyche. *Why did the surveillance fail? Who was the Bank's currency transfer officer? Did the transfer officer betray the bank? Was the Chase Bank explosion a diversion?* Before completing his thoughts, Todd scooted to the back of his chair, placed his head on the desk, and fell asleep.

At 9:00 p.m., Elise and Vanessa checked on Todd, who had not moved since falling asleep. The two women barely managed to move him to the office couch.

Elise later heard Todd milling about the office. She opened the office door, and Vanessa joined her just as Todd slammed the phone. Elise asked, "What's the matter?" Todd became upset when the bank did not answer the telephone. Elise sat beside him, gave him a kiss, and smiled. "Honey, it's Christmas. The bank is closed." Todd apologized and said, "I'm sorry, I'm not with it." Vanessa kissed him and said, "Merry Christmas. We waited for you to wake up before we opened presents." Todd looked at his watch and could not believe the time. "I've never been so mentally exhausted. Looks like I missed Christmas morning. I love you two so much. I'll get it together." Vanessa insisted he first put some food in his stomach. She asked, "When did you last have a meal?" Todd could not remember. He said, "All the days have run together." Both women stood and led Todd by each hand into the kitchen.

Chapter 37:
The Yale-Educated Brain In Action

After Todd ate and the three exchanged Christmas gifts, Vanessa said to Todd, "It's Christmas, and I know you're frustrated about the bank theft. If you don't care to discuss what happened, I'll understand, but I have some thoughts I'd like to share." Sleep and food rejuvenated Todd, and he said, "I would love to discuss it. Perhaps I'm too emotionally involved with this case. I'd appreciate your perspective."

Vanessa began, "Let's be certain I have the correct facts. You said these crooks stole the money without entering the bank." Todd nodded in the affirmative, and Vanessa continued, "If the Federal Reserve never delivered the currency on Tuesday, would there be a magic act? What if the thieves recruited a Bank of America officer? When the bank reopens tomorrow, we agree that we've found a co-conspirator if a bank officer fails to report to work and can't be located." Todd agreed and said, "That was one reason I tried to call the bank this morning."

Vanessa asked, "What are the bank's protocols for receiving money by armored car?" Todd said, "I'm familiar with the process, but each bank may differ. That's the second reason I called the bank this morning. However, armored car protocols are generally the same at each bank. The companies schedule several deliveries during a crew's shift. Some deliveries do not involve currency. They could be delivering non-negotiable paper. There are usually two men in a crew. When delivering currency or negotiable instruments, like bearer bonds, the company notifies the bank's transfer officer of the delivery date and arrival time. When the truck arrives at the bank, one man stays in the truck, and the second man transfers the currency. The bank's transfer officer escorts him to the vault and signs a receipt for the delivery after unsealing and checking the contents of the currency bags."

Vanessa posed a novel theory that Todd had not considered. "What if the armored car, we'll call it the Hollis Armored Car Service, is hijacked, and the delivery is made by a vehicle close to or an exact replica of the Hollis truck or close enough to avoid suspicion. The truck is as fake as the money they deliver. This all works if the bank's transfer officer is a co-conspirator."

Todd saw a flaw in Vanessa's analysis. "How does someone hijack an armored car during daylight with the Hollis guards locked inside an armored truck with weapons and a radio transmitter? All armored car companies contact the truck and the bank's transfer officer if it is late or misses a delivery. If the crew can't be reached or they reply without a password, the company presses a red button at the security desk and pings the armored car's location to the police?"

Vanessa said, "If Hollis calls the bank, they are directed to the transfer officer. The telephone call is meaningless if the transfer officer is a co-conspirator.

"Addressing your other issue in my hypothetical, what if the criminals block the armored car with their vehicle, pop their hood, and pretend to have car problems? Not much time would be needed for another conspirator parked nearby to sneak behind the truck and block the tailpipe, causing the security employees to black out from the carbon monoxide fumes before calling in the delay. A tow truck arrives with two men dressed in Hollis uniforms and winches the in-gear armored truck on a platformed tow truck. They tow the Hollis truck to a distant garage outside the city. Without resistance, the gang blows open the armored door, destroys the electronic tracking equipment, and removes the Hollis crew before they succumb to the deadly gas.

"The currency is removed from the bags, filled with fake money, and transferred to the replica vehicle. Two men drive the replica armored car to Bank of America for the scheduled delivery. While handcuffed and gagged, the Hollis employees recover, only to be injected with a titrated drug to induce sleep for eight hours and are driven and left in a secluded area miles from Salt Lake, with the closest house ten miles away.

"The thieves know they had ample time to leave the country on Tuesday, knowing the missing money would not be discovered until after the Christmas bombing and fire at the Chase bank.

"Hollis only called Bank of America about receiving the delivery and did not bother checking with the other banks on the schedule because there would be no reason if the other deliveries involved non-negotiable securities. It logically follows why Hollis only called Bank of America. I would wager that the Tuesday currency transfer officer is the thieves' co-opted bank employee who assured Hollis that the transfer took place in proper order.

"The relief of the Hollis in-house security officer vanishes when the missing crew fails to return at the end of their shift. On Tuesday evening, December 24th, the thieves were probably in a country with no extradition treaty with the United States when the heist was discovered.

"Honey, what do you think of my theory?" Todd kissed Vanessa and asked if she would like to join the FBI.

Chapter 38:
The New Bride Had It Right

Todd left home for a 9:30 morning meeting with the Bank of America President, Holland Myers. He arrived early but was immediately escorted to the President's office. Holland said, "Agent Montgomery, I've heard spectacular reviews about your work at the Manhattan and Salt Lake agencies." Todd thanked him for the kind words and said, "I'm afraid we dropped the ball on this one, but we will find these men." Both men sipped their coffee, and Todd said, "We believe the bank robbers had an inside man at the bank. When Todd asked if any bank officers were absent today, Holland rose from his chair, frowning, aggravated, and well aware of the questions to follow. Myers said, "Joshua Mathers did not report for work nor call the bank with any reason for his absence."

Todd asked, "Who was the Bank's currency transfer officer for Tuesday's Federal Reserve funds?" Still standing, the President looked at the ceiling. He felt betrayed and said, "Joshua Mathers. He was recently transferred from one of our slower-moving branches for the holiday season. I learned today he requested the assignment. He signed the receipt for the money, knowing the bags contained no currency. We were duped!" Todd told the distraught banker not to be so hard on himself. "Things like this happen. Later today, an agent would like to receive a copy of his file from your HR Department. Lock his office and give no access to anyone or share our conversation. Our forensic team will be here." Todd looked at his watch and said, "They will be here within the hour. We'll be in touch."

***

One of the bank robbers had a brother who was a lieutenant from an organized crime family in Miami. They were based in Chicago, but the outfit had regular operations in South Florida. The brother had a close business relationship with the Cuban government. The high-ranking wise guy arranged the flight to Havana from a private airfield in the Bahamas. Ice Pickle Damato made the introductions. The Cuban officials assured the Harper gang's anonymity and safety, all for one million dollars, with their living quarters in the stately Villa of Saint Isabella.

The Villa of Saint Isabella

D'Amato arranged a celebration at a luxurious restaurant south of Havana. Walking into the dark bamboo restaurant, they enjoyed a cool breeze from the overhead fans. Rex Harper noticed the acknowledgment and respect D'Amato received from the patrons and staff. One of Harper's men said, "Everyone seems to know and like you." D'Amato replied, "What you're seeing is fear. You'll do fine in Cuba. The Cuban officials control the people, and the governing officials are more fearful of our organization than the United States Government."

Chapter 39:
The Canine CIA Informant

Rex Harper wrote a letter to Jessie a week before pulling the Bank of America job. He wrote his nephew's name on an envelope and sealed it. A friend of Harper's, unknown to his nephew, agreed to only hand deliver the message to Jessie after Harper and his crew were safely in Cuba.

A month had passed since the theft when Jessie received his uncle's letter while playing outside with Drifter. Harper wrote, "You must not discuss the existence nor contents of this letter with anyone, not even your aunt. After you've read this, shred the letter and drop its remnants in Salt Lake." He followed his uncle's instructions.

Rex Harper wrote that he had relocated to Cuba but omitted his exact location. He closed by expressing his love and wrote that if Jessie wanted to stay with him in Cuba, he could easily make the arrangements. However, they had to wait a year for investigators to ease up before Harper felt comfortable with the move. He closed, "Please don't try to contact me. The FBI could be intercepting your mail and tapping your phone."

***

Each passing evening, Vanessa and Elise feared they would never see Drifter again. Vanessa, the more vocal of the two women, questioned the reason for the extended absence of their dog. Todd did not want to disclose much detail but finally succumbed to their demands and explained Drifter's extended absence. "Drifter still has important work to do. Agent Adams regularly visits Jessie Harper and takes the dog to our office. Jessie falsely believes Agent Adams is taking Drifter to see her brother and had been forced to find the dog a new home when he married a woman with an extreme allergy to animal dander. We think Harper will try to make some contact with his nephew, and Drifter will be our canine snitch.

"Drifter loves the kid, and the kid loves the dog. Jessie has been living down the street with his aunt. We want and hope Drifter can give us the Harper gang's location. He is well taken care of and loved by the aunt. When he is returned, I cannot break the poor boy's heart. He is a good kid. Agent Adams will have carte blanc when that day arrives to buy any dog Jessie wants." Vanessa and Elise became emotional with Todd's concern for Jessie. They both wiped their tears and kissed him, and Vanessa said, "Elise, we are so lucky to have such a caring man as Todd."

***

When the time arrived for Drifter's pretend visit with her brother, Agent Adams drove straight to the downtown FBI office. The area still smelled from the explosion and fire at the Chase bank. When Agent Adams arrived, Todd and Drifter took a ride around town. Todd still could not make any logical sense of debriefing a dog.

He asked, "Well, old boy, any news?" With Drifter's affirmative response, Todd hit the brakes and pulled to the side of the road. Drifter disclosed, "Jessie and I were in his room, and he was reading about Cuba on his computer and talking to himself. His Uncle Rex and the gang are in Cuba. Some guy I had never seen before delivered a sealed envelope to Jessie. I could tell he didn't know the messenger."

Todd asked if Drifter could snatch the letter for him to read and have the lab analyze it for prints. Drifter sadly broke the bad news. "Jessie shredded the letter and scattered the pieces into Salt Lake." Todd returned to the office and gave Drifter to Agent Adams to return the dog to Jessie.

***

As a seasoned planner and thinker, Uncle Rex didn't overlook the realities of international relations, particularly with a dictatorship like Cuba. Although the country did not have an extradition treaty with the United States, the American bandits seeking safe refuge from justice became less secure in 1977, when the U.S. Government established limited diplomatic relations with the outlaw country. Cuba's decision to extradite some but not all American criminals became the Cuban Government's new revenue source. Those exiled in Cuba without substantial 'sponsorship' and refused or had an insufficient tribute for Cuba's hospitality would soon find themselves jailed, shackled, and returned home to answer for their crimes.

Rex Harper neutralized this risk and felt secure by the sponsorship of America's organized crime families of New York, New Jersey, and particularly the City of Chicago. The Cubans knew the protections of international law did not protect them from the Mob, which operated by its own rules. Assassinations and political disruptions were a real risk, and the Cuban government feared the consequences if they reneged on such powerful sponsorship agreements. International Law offered the poor island country only limited protection from the mighty power of the United States after losing the country's powerful protector—the Soviet Union, evidenced by the Bay of Pigs fiasco during the Kennedy Presidency.

Chapter 40:

Who Has The Stronger Protector?

On December 30th, Todd arrived home from work at 6:00. He had enough of the late-night sessions. The Rex Harper gang wasn't going anywhere.

After a light dinner, they packed for the New Year's Day marriage of Vanessa and Todd, followed by a honeymoon and a ski trip at the Yellowstone Ski Club.

Elise and Vanessa asked about the Harper case. Todd said, "We know he and his gang are somewhere in Cuba. The country has no Extradition Treaty with the United States, but all hope is not lost. We believe the Harper men have ties to the American Mob. With the Mob as their benefactor, our job is more complex but possible. We used rendition in the past, and perhaps we can use it here. The bank thieves have their connection with the Mob, but I have my connection with the CIA.

***

Todd's reference to rendition originated in America during the late 1800s, and there were virtually no legal restrictions, only domestic and international political considerations for seizing U.S. citizens based upon probable cause. For the

Harper crew, the United States had to establish the commission of a serious or significant action to commit such a crime. Judge Proud's issuance of arrest warrants settled the probable cause requirement.

The mass planning and the mounds of paperwork to obtain authorization for rendition could wait until Todd returned from his one-week vacation. During his absence, CIA operatives in Cuba would work to ascertain the location of the Harper men and the banking depository for the money, most likely in an off-shore numbered bank account.

Chapter 41:

Happy New Year

More than a year had passed since Irving L. Muckelman died in the tractor accident at his Sherburne, New York, farm. The sad but bizarre events were not lost on Todd Montgomery. He thought about Mucky and remembered his promise to include Drifter on their one-week trip. He had Agent Adams call Jessie to explain that her brother would like to have the dog for the week of his Mother's visit to Salt Lake City. Jesse agreed, and come New Year's morning, Todd picked up Drifter and assumed the mythical brother's identity. Neither the suspects, Jessie, nor his aunt had ever met the mythical brother or Todd. The visit provided the FBI Assistant Director with an opportunity to see the layout of the aunt's house.

***

When Todd arrived home, Vanessa and Elise fawned over Drifter. Elise said, "Look, it's the famous undercover FBI dog." Drifter enjoyed the attention and said, "I was afraid you guys forgot me. I'm glad to be home. The aunt's place smelled like an old person's home."

Todd loaded the Rolls, and they began their 375-mile journey with a brief detour to Jackson, Wyoming, for the wedding of Todd and Vanessa by an old friend of Todd's, a soon-to-be-retired Circuit Court Judge. Wyoming did not have a marriage waiting period.

Driving on dry pavement, with a speed limit of 80, Todd felt comfortable driving 100 most of the way north on Interstate 15 to the Yellowstone Club. After settling into their house, like magic, heavy snow blanketed the ground. Todd, his twin wives, and Drifter enjoyed the view from a roaring fire.

The Yellowstone Ski Club

***

The fundamentalist protocol for multiple marriages did not allow the husband and the wives to sleep together. The family prepared a monthly schedule of each wife's day and evening with the husband. Since Todd had two wives, the time allocation did not present any problem. However, the scheduling became quite challenging for a fundamentalist husband with six or more wives. Thanks to Drifter, Elise and Vanessa never slept alone.

***

As much as he tried, Todd could not leave the Harper case behind. He ran various scenarios, its prospects of success, and probable domestic and foreign consequences. He had filled a legal pad with notes by the end of their week's stay.

Most evenings, after promising Drifter a doggie bag, they ate dinner at the Club's restaurant. They became quite an attraction among the other members. When the Montgomery twins entered the restaurant, a male diner's wife caught him ogling the twins. Under his breath, he said, "Now that man knows how to live." The perturbed wife told her husband to quit staring. "Keep your eyes on your plate. You're spilling gravy on your tie." Unlike Elise, Vanessa loved the attention and seemed to encourage such interest with her seductive attire.

Vanessa Montgomery

Chapter 42:

A Bad Bologna Sandwich

A day after returning to Salt Lake City, Todd returned Drifter to Jessie and later traveled to Langly, Virginia, to see a longtime associate at the CIA. He had met John Statler while assigned as the Manhattan Director of the FBI. Todd explained the status of the Federal Reserve's twenty-three million dollar currency theft and wanted to know the agency's position for the involuntary rendition of Rex Harper and his men from Cuba.

John Statler smiled and said, "I already know about your case. The CIA is always ahead of the curve. We have no written policy. We examine the reasons for initiating the action, the logistical odds of succeeding, and future domestic and international consequences.

"CIA intel knows Harper paid Cuba one million dollars, and who knows what sponsorship money the Mob paid. The Cuban Government is corrupt but not stupid. They are well aware that if they reneged on their sponsorship agreement by implicitly allowing the involuntary rendition by the United States, they would have a Hobson's choice. After arranging a surprise ninety-mile boat ride to Havana, the Mob would reclaim their lost respect by causing Cuba to suffer domestic instability and the likely assassination of the Castro brothers. If they expressly opposed rendition, they would suffer the wrath and power of the United States military." It seems counterintuitive, but Cuba is more fearful of the American

Mafia than the United States, because the Mob is not constrained by international law or condemnation by other nations."

Todd said, "With the need to replace the lost income from the Soviets, Cuba's plan to think outside the revenue box backfired. They have been paradoxically placed in the unenviable position of finding themselves in an inescapable trap."

Todd suggested a solution. "I understand the CIA isn't in the business of telegraphing their moves, but what would be the likely outcome if CIA and FBI operatives leaked a bogus involuntary rendition? We force-feed Cuba a rancid bologna sandwich, something that looks good but tastes bad. The Mob believes our government would not consider involuntary rendition without Cuba's tacit approval." John liked the idea and said, "I'll see what the big boys think." They shook hands on the plan, and Todd said, "There is no rush to decide. I'll fly home tonight." Call when they decide."

***

When Todd arrived home, the family asked what had been decided. He provided limited information. "We exchanged ideas, but most importantly, the CIA was on board. We shouldn't have any problems with the involuntary seizure of the Harper group in operational terms or overstepping

International Law. When we next meet at Langley, the FBI, the CIA, and the IRS will hopefully have the name and location of Harper's off-shore bank.

"Designing a plan outside the box is not Cuba's strong suit. The feeble-minded Cuban Government fell into a paradoxical trap of their own making. If Cuba even hints at breaking its pledge to the Mafia, the Mob would consider this an egregious breach of trust and a clear sign of disrespect. Cuba's Hotel bookings would increase because many angry American-Italian Mobsters would descend upon the corrupt island.

"Let's stay home for dinner and order delivery tonight. I'd like to just relax this evening and watch a movie." Elise didn't care what they did. It was her night to be with Todd.

Chapter 43:

Bingo

Todd Montgomery returned to Langley on Monday, January 13th. Gazing out the jet's window, he believed 2014 couldn't have started any better. His positive attitude and confidence had returned, in sharp contrast to his dour mood after the embarrassment from the Federal Reserve heist. He firmly believed the shrewd plan he devised, if accepted by the CIA, would soon have Havana and the Mob at each other's throats. His marriage to Vanessa enhanced his personal life as well.

The follow-up meeting with John Statler did not last long. The CIA endorsed Todd's recommendation. Like Mr. Muckelman often said, Todd's plan 'killed two birds with one stone.'

The CIA had previously consulted and convinced the Defense Department to schedule naval war games off the Cuban coast. The Department of Transportation grounded all Cuban flights from the United States, and our South American allies followed suit. The CIA disclosed little detail to our allies in the southern continent, explaining the secretive operational plans of the United States involved national security in the region.

Todd and John knew that regardless of a poor country's political sentiment toward the United States, a pledge of secrecy of operational military movements was typically disclosed to all bidders.

Two days after the United States initiated the scheme, Cuba panicked and declared martial law after rioting in the streets of Havana. The Cuban government knew the reason behind the upheaval, and the U.S. Mob also closely monitored the situation.

***

With the United States District Court subpoenas, the IRS's phishing expedition located Harper's offshore numbered bank account. The Grand Cayman Island Depository, a shell loophole bank permitted by the Foreign Account Tax Compliance Act, allowed these offshore banks to accept funds from United States citizens and falsely self-certify the depository information to the Internal Revenue Service.

Billions of tax dollars were evaded because the underfunded IRS approved the bank's self-certification through its portal without any meaningful investigation and a de minimis likelihood of a later bank audit. Because of this, the odds of locating the Harper gang's account equaled the odds of hitting the correct numbers and color on a Vegas Roulet Wheel.

After a net asset audit of the third bank on the IRS list, the U.S. investigators had a good day at the casino and found the numbered account holding the stolen Federal Reserve currency. The Bank's officers were returned to the United States wearing handcuffs and leg chains, courtesy of the FBI.

While other countries speculated upon the next move of the United States, the question on the streets of Havana was not whether the U. S. Military would commence involuntary rendition. Instead, the question was when the United States would invade Cuba after the corrupt island nation denied voluntary rendition caused by the country's horror from the Mob. The CIA and FBI watched as Cuba took the first bite of the rancid sandwich. Todd's U.S. ploy succeeded without the risks associated with invasion. Two days passed when reports confirmed six American outlaws were found dead in The Villa of Saint Isabella. The bodies were returned to Miami, and the autopsies confirmed the Harper gang had been eliminated. The Cubans maintained the gunshot wounds were self-inflicted, but the United States and the Mob knew better.

Our government recovered all the stolen money, less the one-and-a-half million dollars of sponsorship money paid to Cuba by the Mob. The United States Mob money was later recovered when Brazil froze Cuba's assets.

The Mob still had unfinished business with the Cubans for violating their pledge of a safe harbor to a client. Retribution would help to erase the embarrassment of the broken pledge when the Cubans murdered the Harper gang.

Italians arrived in Havanna a month later and registered at the posh beach hotels. The desk clerk asked an American-Italian guest if he had traveled to Cuba for business or pleasure. With a gravel voice, the gruff man from New Jersey blew a large plume of cigar smoke into the face of the desk clerk as he answered his question with a provocative answer. The rotund guest frowned and said, "Both. It's business when I shoot someone, and seeing them fall to the ground is pure pleasure."

The only fallout for the United States occurred on Wall Street. The sound of war drums spooked the market. However, the exchanges erased their losses within a week.

Chapter 44:

Victories Can Be Sweet and Sour

During his flight home, Todd worried about Jessie losing his uncle and dog. He worried the two events might be too much for the youngster to process. All the twenty-four-seven newscasts on cable bombarded him with the news of his Uncle's death.

Todd decided Agent Adams, and he would meet with Jessie to console him for his loss. He also decided to seek some outside help from a friend. On January 23rd, Todd telephoned George Markarian in Florida.

George welcomed Todd home to Salt Lake City and congratulated him on the elaborate scheme he had accomplished. Todd said, "That is the reason I called. Max Harper was the appointed Guardian for his nephew, Jessie Harper. They were very close, and I'm ashamed to say the FBI used Jessie to obtain information about his uncle. He is a decent and smart boy, but the family is dirt poor. I imagine little money has been set aside for his education. Could the foundation help him with a scholarship? He is very bright, and I will gather his school records. Could Jessie and I come out for a few days so you can meet and evaluate the boy? If you determine he is underserving, I'll understand. I'm just trying to streamline the process. We'll stay at The Longboat Key Resort." George interrupted, "Bullshit! There you go again! You'll stay at our place! I may need more help with my speeding tickets."

***

Agent Adams and Todd took Jessie to lunch at the Copper Onion. When they arrived at his aunt's house, he asked, "Where's Drifter?' Todd considered what he would say to the youngster and decided the truth would be brutal, but the time for honesty had arrived.

While driving to the restaurant, Todd began his confessional to Jessie. "Agent Adams and I work for the FBI. Drifter is my dog, who was specially wired to learn about your uncle's plan to steal money being transferred from the Federal Reserve Bank to the downtown Bank of America. Agent Adams has received permission to allow you to buy any dog you choose. I am so sorry we misled you. We both think you're a wonderful boy and hated deceiving you, but we were only doing our job. We are both sorry for your Uncle Rex's death. We never pointed a gun at him. The Cubans killed him and all his partners."

Jessie hesitated with his response and said, "I figured Uncle Rex didn't have a real job, and when I tried to talk to him about it, he always became angry and left the room. I know stealing is bad, and I understand you guys were doing your job to protect the country. I am not mad." They turned into the restaurant parking lot, with the heavy weight lifted from Todd.

While eating lunch, the subject turned to academics. Todd asked Jessie if he wanted to go to college. The boy did not hesitate. "I sure do, but I'm afraid I won't be able to afford it. I'm in the 8th grade, so I have some time, but I really want to go and make something of myself." Todd knew the boy was eleven and asked, "Shouldn't an eleven-year-old be in the sixth grade." Jessie explained he skipped two years because his teachers believed he wasn't being challenged. He said, "I guess they were right because I still make all A's."

Todd told Jessie about George Markarian and his position as the Director of the Muckelman Scholarship Foundation. Todd said, "If your aunt approves, how would you like to visit Mr. Markarian in Longboat Key, Florida, and discuss your wish to attend college?" Jessie jumped at the offer but asked, "Why are you doing this for me?" Todd gave a simple answer. "Everyone deserves an opportunity, and your academic standing screams for the chance to attain your dream."

Jessie asked if he could see Drifter after lunch. Todd returned Agent Adams to the office, and while there, Jessie called his aunt for permission. The youngster surprised Todd when he said, "I knew she would say yes because she is tired of having me around. She only gets social security, and I overheard her tell our neighbor that having me permanently had become an expense she couldn't afford." Todd blamed himself for placing the young boy in such an untenable position.

***

Vanessa and Elise were surprised when Todd arrived home with his guest. Drifter and Jessie loved their reunion as they rolled on the floor in a playful wrestling match. Todd asked Elise and Vanessa to come into the kitchen while Jessie played on the floor with the dog. Todd said, "Jessie took the news of Drifter and me tracking down his uncle in a mature way. He said, 'We were protecting the United States.' Everyone knows the Cubans murdered the men. The United States never set foot in Cuba. The man I worked with at CIA operations pledged this to me. But in a way, I feel somewhat responsible for the uncle's death. Stealing money didn't justify the brutal murder in Cuba. During our lunch Jessie said he had overheard his aunt mournfully tell a neighbor she could not financially afford to keep him permanently. I feel awful. He said his aunt and uncle were his only relatives. What can we do?"

Chapter 45

A New Life

When Todd drove Jessie home, he asked if his aunt had time to speak with him. He wanted to talk to her about what Elise, Vanessa, and he discussed in the kitchen. When Aunt Dorris appeared in the living room, Jessie and Drifter left to play outside.

He introduced himself and said, "I want you to know at lunch, Ms. Adams and I told Jessie we work for the FBI, and we arranged to wire Drifter and place him in his uncle's house to learn about their plans to steal a currency delivery by armored car to the Bank of America. Jessie is an exceptional boy. He is smart, quite mature, and understood we were doing our jobs. He believed the FBI protected Americans. I told him the United States Government had nothing to do with the murders in Cuba. The murders were not justified, and our Government did not participate or set foot on Cuban soil. Before I go further, I would appreciate your thoughts about what I told you."

Dorris smiled and said she had no ill feelings for what had happened. "I have always told Jessie that respecting law and order and rules of civilized conduct makes this country great. Max never grew up, and I thank the Lord Jessie did not turn out like his uncle."

Relieved about the aunt's attitude, Todd continued, "I asked Jessie how he liked school and about the grades he received. He surprised me when he said he was in the eighth grade. I said an eleven-year-old should be in the sixth grade. But he explained his school felt he wasn't challenged at that grade level and advanced him by two years. Even with the advancement, Jessie said he still makes all A's." This surprised Dorris because Max never shared Jessie's progress in school. Todd continued, "Jessie wants to attend college but said the family could not afford the cost. Please understand we were not prying when Jessie volunteered that he overhead you tell a neighbor you could not afford to keep him permanently."

Dorris grabbed a handkerchief and began to cry. Todd tried to console her, "You should not be ashamed. Things in life happen without the need to assess fault. Many people are in similar situations. Caring Christians help those in need, and I propose to do the same. I have discussed Jessie's schooling with my wife and her sister, who lives with us. I can afford what Jessie needs because of an inheritance. However, the FBI has strict conflict of interest rules preventing the direct use of my private funds. I took the liberty of calling a close friend who lives in Florida and is the Executive Director of an Educational Foundation. I told him about Jessie, and he is eager to meet with him. Jessie looks forward to the meeting, and I plan to accompany him.

"I know you've received a load of information to process. Think about all of this and talk with Jessie. Here's my card, and these are my FBI credentials. Your nephew will be in good hands. Call me in a few days." Dorris hugged Todd and said, "God bless you."

***

Todd felt wonderful during his drive home. He stopped at a Mormon Church and prayed for success to transform a tragic event into a miraculous chance for a deserving boy who had received no breaks during his short life.

He called George from his office. "Yes, George, you heard right. Jessie is eleven, has advanced two grades, and continues to make all A's. The boy is smart and quite mature for his age. His parents abandoned him and left him to live with his uncle, the leader of the gang murdered in Cuba. Jessie's aunt is a sweet woman who wants to do more for the boy but lacks the financial ability to care for him. I am thinking of giving his Aunt Dorris a few days before asking if he can stay with us a week before we fly to your place. I'd like us to know each other better." George said he had some ideas and suggested they fly out on Saturday, February 1st. He said, "Maria will be in New York. We'll have the place to ourselves."

***

During Jessie's week-long visit, Todd asked Elise and Vanessa what they thought of Jessie. Both women shared a common opinion. Vanessa said, "While you were out, we felt the room light up as he played with Drifter. You were right. He loves that dog. It seems everyone agrees with the methodical way you're handling this, culminating with our visit to Florida." Todd laughed, saying, "Are you telling me you two plan to join us?" Vanessa confirmed Todd's conclusion. "I always said you were sharp. You bet. Nothing better than some South Florida sun in February." Drifter demanded, "I'm going too!"

Chapter 46:

George's New Friend

Elise and Vanessa had instructed Drifter not to talk in front of Jessie. Todd did not want to overload the boy on his first day. When Todd arrived home with Jessie, he excused himself to check his messages in the office.

For the first time, Jessie commented on the beautiful sisters. He said, "You two are identical and super pretty. Which one is married to Todd? Under her breath, Vanessa said to Elise, "You've got seniority. Tonight is your turn anyway." Elise claimed the honor, and the inquisitive youngster asked, "Gosh, how does Todd tell you two apart?" Before either could answer, Todd returned, heard the question, and answered, "Jessie, the women are visibly identical in appearance, but their personalities could not contrast more. Vanessa is the free spirit of the family, and Elise is grounded on the conservative side. Once you know them, it is pretty easy to tell them apart."

Todd appeared older than his age and had begun his growth spurt. He looked like a high school student and had the athletic physique of a basketball player. He rarely socialized with his peers because of his excessive shyness, which wasted his handsome features with women.

***

The travelers arrived at the FBI hangar. Todd decided to fly privately and reimburse the agency for the cost of travel. The passengers, Todd, Jessie, Elise, Vanessa, and Drifter, boarded the jet for the long cross-country trip from Salt Lake City to George's Longboat Beach home, perched on the Western edge of the Gulf of Mexico. Jessie's excitement about his first airplane flight grew as the plane taxied for take-off. He again thanked everyone for their kindness. He looked at Drifter and hugged the dog. Todd leaned back into his plane seat and chuckled to Vannessa. She asked what was so funny. Todd explained, "You'll learn when you meet George."

***

They weren't flying in George's Citation X, but because of the strong tailwinds, they arrived in Sarasota earlier than planned. When he told George of the earlier arrival time, he asked if George would bring Maria's van. George asked, "Old buddy, did you bring the entire BYU football team?"

As the government jet taxied to the hangar, Todd had given clearance to allow George to drive the van onto the tarmac. George had a big smile and a Texas-size wave when he saw Todd, but the wave slowed when he saw Vanessa and Elise.

After the entourage exited the plane, Todd introduced Jessie to George, who said, "I've heard wonderful things about you. Do you have a favorite subject in school besides coeds?" George could not take his eyes off the two women. While talking to Jessie, he positioned himself to view Utah's twin peaks. Jessie said, "I love to learn anything new. I have an equal interest in all my courses."

Drifter whispered to Todd, "It's hilarious to watch George try to figure out which one is Elise so he can flirt with the right woman." Todd agreed with Drifter, and he clued the gals into a Markarian-style prank to have Drifter return to the airplane.

Before George could question Todd which of the twins was his wife, Todd said, "Girls we better get Drifter off the plane before he destroys it." Again disappointed, George watched the two women follow Todd with a slow, exaggerated, and salacious walk into the airplane.

After unpacking, Todd found George and grinned, while he crafted an insincere apology for failing to introduce Vanessa. George would not relent and asked him, "Old buddy, how can you tell the women apart?" Before Todd could respond, Jessie walked into the living room with his transcripts and handed them to George. Todd excused himself, left the room, and chuckled with his wives. Vanessa said, "Jessie followed your directions well. Your new accomplice will fit in quite well with this goofy family."

***

George returned to earth and looked at Jessie's transcripts and saw the boy had aced all of his courses. He said, "Jessie, the Foundation is intended for college, but its general purpose clause will allow us to use foundation money for the best prep schools in the Northeast. How do you perform on standardized testing?" Jessie said, "I have no idea. The school never had us take one." George told him not to worry, "I'm sure you'll do fine. I'll pull the Foundation documents and take a closer look, and I'll be up to speed tomorrow."

***

George planned to rise early the following morning to see which twin walked out of George's bedroom. He overslept, but his plan had been doomed from the start because, for the first time, Todd had slept with both wives during his Longboat visit, and for the first time since joining the family, Drifter slept alone.

The following evening before bed, George set his alarm clock for 7:00 a.m. and began his vigil on the living room couch, only to be shocked to see Todd leave his bedroom with both women. Todd had outsmarted George and told him, "I hope you slept as well as I did. Your king-size bed is quite accommodating." Foiled again, George muttered, "Damn!"

***

Elise, Vanessa, and their prep chef, Jessie, made breakfast. Todd and George sat on the couch, avoiding any mention of the threesome's departure from the bedroom.

After an awkward pause, Todd said, "I don't believe Aunt Dorris had the opportunity to be named as Jessie's Guardian. Without an appointment order from the court, she has no legal authority to make any decisions upon his behalf." George, who had regained his composure, proposed an idea that stunned Todd. "You said Dorris could not afford to take care of him. Even if she were appointed, that would not resolve the problem. Why don't you adopt him? It's obvious he loves you all."

After breakfast, George and Jessie met alone in the library. They conferred for over an hour. Afterward, George told Todd, "This kid is advanced. I bet Jessie could keep up with college students. There is a great prep school in New York that would be ideal. "The Avenues of New York" is in Manhattan's Chelsea Neighborhood with a 70,000-square-foot campus. The Foundation has dealt with their admissions office before, and they are good people with a sincere concern for their students. With their advanced placement curriculum, he could graduate early and get a jump start at some of the country's finest universities. If my evaluation of his abilities is correct, we'll need to shuffle some paperwork, and he is in." Todd said, "George, what I like about you is you're a man of action. Well, the ball is in my court now." George still pestered Todd about the twins' identity for the remainder of their visit.

Chapter 47:
The Ultimate Paradox

While returning to Salt Lake City, Todd considered George's suggestion about adopting Jessie. The resolution of the issue depended upon the decision of five people. Todd had a complicated judgment to make. If he chose to adopt, it would have to be for the right reason. His motivation could not be a selfish wish to eliminate his unfounded guilt of causing Rex Harper's death. Did he love Jessie as a son? At the end of the four-hour flight, he came to a decision. Before unpacking, he called Aunt Dorris when they arrived home and wanted to talk with her that evening.

***

 While Jessie and Drifter played outside, Todd affectionately took the hand of each wife, led them to the office, and shut the door. He had not told them about George's adoption suggestion.

"Girls, I want to talk with you about Jessie. George and I had a long discussion, not only about Jessie's education but his legal status as well. There is no guardianship. Legally, he is in limbo with no adult lawfully capable of making important decisions for him without petitioning the court. We know his aunt loves him but cannot afford to take care of his needs as a parent should. Suppose the court determines his aunt doesn't qualify as a suitable guardian. Without any other known relatives to assume such a responsibility, Jessie would become a ward of the court and end up in foster care. I know that is not what anyone wants.

I want to adopt Jessie, but we all must agree as a family. I am seeing his Aunt Dorris this evening to discuss the good news about his schooling and the legal issues involved. Our decision to adopt that wonderful little boy cannot be my decision; it must be our decision." Vanessa and Elise rose from their couch, kissed Todd, and Elise said, "Honey, when did you learn to read minds? We would love to have Jessie as our son." The two remaining parties to decide were five minutes away. As the snow began to stick to the road, Todd left to see Aunt Doris and Jessie.

***

When Todd spoke to Jessie and his Aunt, he began with the news of the Foundation underwriting his prep school enrollment. "It's an old and respected prep school. If Jessie does well, he can choose among the best universities in the country. However, when I spoke to George Markarian, the Foundation's Executive Director, we discussed an important legal issue that needed resolution. There is no Guardian appointed for Jessie. I remember discussing your financial concern about raising your nephew as a parent should." Dorris said, "Todd, I love Jessie and want the best for him. When he walked over with Drifter that evening, we assumed it would be a short visit until his uncle returned home."

Todd continued, "George is aware of your problem; no other family members exist or at least have not come forward. In such cases, the Children and Family Services could recommend that the court place Jessie in foster care." Jessie jumped out of his chair and shouted, "No way will I live with strangers!" Todd asked, "Jessie, how would you and Aunt Dorris feel if we adopted you? You would become part of our family. We all love you, and we are family-oriented. Our hearts welcome you both as family with open arms. The Montgomery family increased from three to six when Drifter became the first to agree.

***

As Jessie's school year approached, a menacing black cloud followed the Montgomery household. The twenty-two hundred miles between Salt Lake City and Manhattan seemed far removed from the uncomplicated lifestyle and culture of Salt Lake City.

Kismet struck again and dispersed the cloud cover when the Director of the Manhattan FBI Office resigned for a position in the private sector. After a family council meeting, they unanimously decided that Todd should return to his old office. The former issue of Todd and Elise's loneliness of living in New York City became a moot issue because Todd and Elise now had a real family, and Jessie's school would be a subway stop from their son's new home.

***

Todd flew to New York ahead of the family and wanted to settle into his new office and explore suitable housing. Sitting alone on the airplane, he wondered whether he had just experienced the happy side of a unique paradox caused by a unique series of events: the embarrassment of being hoodwinked by the theft of the Federal Reserve currency, his success of turning Cuba, the Mob, and the Harper gang, once a troika of an evil alliance, into explosive adversaries, the murder of the man whose family he adopted, and a possible Ivy league education for Harper's nephew. The sequence of these events was improbable, but because of Todd's moral compass, as expressed on his monogrammed wedding ring, and with the good Lord's help, they were transformed into a celebrated ending.

Chapter 48:

Epilogue

During the February term, the United States Court of Appeals Chief Judge for the Second Circuit of New York swore in Todd Montgomery at the downtown Federal Building as the new Director of the Manhattan FBI office. The event meant more to Todd than his prior ceremonies because he had his new family watching from the visitor's gallery.

The ceremony and courtroom impressed Todd's son. When in session, three massive leather chairs were the focal point for a panel of three randomly chosen Appellate Judges who would hear attorneys argue their respective positions concerning the soundness of the lower court's ruling. The magnanimous courtroom ushered the respect and awe inspired by one of the three independent branches of our Constitutional Government.

***

After the swearing-in ceremony, Todd's family and George Markarian enjoyed lunch at the posh and historic Sparks restaurant on the East Side of Central Park. Patrons loved the restaurant not only for its incredible food but also for observing patrons, who were famous entertainers and well-known politicians.

Todd shared a notorious historical event that took place at Starks. In 1985, the Gambino crime family boss Paul Castellano and underboss Thomas Bilotti were gunned down outside the entrance to the restaurant; Mobster John Gotti ordered the hit. Todd said, "With the hit, Gotti ascended to the head of the Gambino crime family. After three acquittals in the 1980s, the flashy Mobster's life of crime ended with his 1992 conviction and his 2010 cancer death while incarcerated." George added, "Gotti, known for his flashy attire, purchased his handmade silk suits from Cushman's Clothing. Our old buddy, Irving L. Muckelman, always waited on him."

***

After Lunch, Todd drove everyone to the Trump Tower. He did not know George Markarian had secretly facilitated his family's move to the spacious complex by using his influence to place his old buddy's name at the top of a long waiting list.

George would return to Manhattan in the spring, and both men would develop a close relationship. Before leaving for the office, Todd made a few telephone calls from the restaurant.

While checking his mail at the Manhattan office, he saw a memorandum from the United States Attorney General, Maxwell Hart, appointing him to chair a Select Congressional Subcommittee to investigate President John F. Kennedy's 1963 assassination.

Since the horrific event, the Attorney General believed in a conspiracy theory behind the murder of the President. He and members of Congress on both sides of the aisle criticized The Kennedy Act of 1992, which promulgated that no records of the assassination could be withheld without the agreement of the Assassination Record Review Board. The statute provided that if The Review Board ordered the release of records, the President would have an absolute veto of the Board's decision without the recourse of an override vote by Congress.

Todd believed the statute violated the separation of powers clause of the Constitution and felt a Supreme Court challenge, not a Special Congressional Select Subcommittee, should be the proper course of action. Since 1992, every President has vetoed the Review Board's decision to release the records. Todd Montgomery feared an

investigation would open a nasty can of worms. He asked his
secretary to get the Attorney General on the telephone.